Disaster Recovery Planning

For Computers and Communication Resources

Jon William Toigo

John Wiley & Sons, Inc.
New York • Chichester • Brisbane • Toronto • Singapore

Publisher: Katherine Schowalter
Editor: Robert Elliott
Managing Editor: Robert S. Aronds
Text Design & Composition: Jo-Ann Campbell, *mle design*

This text is printed on acid-free paper.

REQUIREMENTS:

An IBM® PC family computer or compatible computer
with 256K minimum memory, a 3.5" high-density floppy drive,
PC DOS, MS DOS, or DR DOS Version 2.0 or later, and a printer.

This publication is designed to provide accurate and
authoritative information in regard to the subject
matter covered. It is sold with the understanding that
the publisher is not engaged in rendering legal, accounting,
or other professional services. If legal advice or other
expert assistance is required, the services of a competent
professional person should be sought.

Library of Congress Cataloging-in-Publication Data

ISBN 0-471-12175-4 (3.5")

Printed in the United States of America

10 9 8 7 6 5 4 3 2 1

*This book is dedicated to Margaret Romao Toigo,
without whose patience and assistance
this project might not have been completed, and also
to all disaster recovery planners who
do not regard themselves as gurus,
because most of them are.*

Contents

CHAPTER 9 **NETWORK AND COMMUNICATIONS RECOVERY PLANNING** **217**

ACKNOWLEDGMENTS

The author wishes to acknowledge the following persons for their assistance and encouragement over the course of this project:

Craig Jensen, Frank McEntire, Roger Foresberg, Milt Maughan, and Charlie Fox of the Association of Contingency Planners of Utah, Salt Lake City, Utah, who repeatedly find interesting things for me to do and interesting problems to solve.

Rich Mansfield, of Publishers Resources Inc., Metuchen, New Jersey, who was part of the original conspiracy.

David Brack, The Brack Group, Safety Harbor, Florida, and Ed Yourdon, New York, New York, who introduced me to the world of data-flow diagramming.

Kyle E. Jones, lately of Technologically Innovative Processing, St. Petersburg, Florida, in the hopes that his vision will be realized and become something worthy of disaster recovery planning.

The many readers of *Disaster Recovery Planning: Managing Risk and Catastrophe in Information Systems* who asked that this detailed treatment of the subject be written.

My children, Alexandra and Maximilian, and my family and friends, for their warmth and reassurance.

Finally, to Bob Elliott, my editor, who has the patience of Job.

Purpose and Scope of this Book

PURPOSE OF THIS GUIDE

Today, more than at any previous time, computers provide mission-critical information services for the modern corporation. Industry surveys show that the average corporation spends between three and seven percent of its annual revenues acquiring hardware and software for corporate data processing. They pay another two to five percent for maintenance contracts on this investment. And the preponderance of the 20 to 30 percent of revenues that companies pay to train their employees is aimed at equipping users to make the most of automated systems.

These statistics suggest a modern truism: Companies depend on their computer systems to make them competitive in the marketplace. They rely on the *efficiency* of the computer to reduce labor costs and to improve methods of information collection, processing, and distribution. Corollaries to efficiency are the values of *access* and *integrity*. Companies need to assure that their systems and networks will be available for use when needed and that the information manipulated by systems and networks will be reliable and free from error.

As straightforward as this may seem, the dependence of the company on its systems and networks is rarely considered from the perspective of the vulnerability that dependence creates. The reason is fundamental. Most corporate cultures are aggressive and positive. Businesses are driven by a desire for gain. Monies spent are justified by the gains that are realized, typically measured in higher market share ratings or cost-performance improvements. In a corporate culture thus motivated, the consideration of negative potentials is rarely perceived as valuable.

Still, there is a growing body of empirical and statistical data to make the case that senior management must consider the possibility of a disaster and prepare for it. Industry studies have depicted the ramifications of a disaster for business activities in stark terms. These studies suggest that companies would sustain critical or total loss of their ability to conduct business within 15 days of a data processing outage and would have less than a 25 percent chance of ever recovering.

Given these alarming findings and the extensive coverage given to recent man-made and natural disasters in the news media, it is surprising that many companies continue to ignore the possibility of disaster and fail to plan for their recovery. Repeated case studies support the idea that a contingency plan—a plan for reacting to and recovering from a disaster—both enhances the survivability of the company and reduces its susceptibility to avoidable disasters. Yet, less than half of the companies involved in national studies conducted over the past decade report having a disaster recovery plan.

The avoidance of disaster recovery planning is the result of several dynamics. At a psychological level, it can be argued that we tend to avoid looking at the unpleasant possibility of a disaster. Also, for many years, we contented ourselves to rely upon business interruption insurance and special data processing coverages as a hedge against a system glitch.

Added to these factors was the time-honored calculation of cost-efficiency. Insurers suggested that the likelihood of a disaster affecting the corporate data center was less than one percent. Expending capital and other resources to develop a recovery capability that would probably never be invoked was simply not cost-justifiable for many companies.

Moreover, computer hardware vendors have made much of the purported invulnerability of their computers, storage devices, and telecommunications networks which have only now, in the wake of some very high-profile disasters, begun to be called into question. Even today, however, it is not uncommon to read about so-called "sealed disk drives" that are impervious to contamination and crashing.

With so much against contingency planning, what can a book such as this one hope to accomplish? The answer to this question is three-fold. First, this book seeks to set the record straight. With all of the misinformation and outright disinformation about the principles and techniques of disaster recovery planning, there is a need for a document that treats the subject objectively, realistically, and accurately.

Thus, this book is offered as a baseline to aid the reader in culling through the literature of disaster recovery planning, especially the marketing materials of vendors and consultants, and to separate the grain from the chaff. Moreover, it is hoped that this book will help readers develop a more informed and critical view of contingency planning. Such an outlook will serve them both in evaluating the products and services of the disaster recovery planning industry and in preparing an effective disaster recovery capability.

The second reason for this book is to offer a model for the planning process that can be emulated by readers in their own planning projects. At present, there are no fewer than thirty contingency planning methodologies for sale by vendors who portray their products as disaster recovery *solutions*. Users of these so-called *canned plans* are asked to believe that disaster recovery planning can be as simple an endeavor as adding their particular information to a boilerplate document. Print out the document, bind it, and—*Voila!*—instant plan.

In contrast to this simplistic approach, many disaster recovery consultants continue to portray their methods as rarified, enigmatic, and somehow beyond the reach of the novice. While they correctly point to the fact that most plans initiated by in-house personnel are never completed, they suggest incorrectly that this is the result of a lack of ability on the part of the planner. More often, failure to complete a plan results from a loss of management support for the endeavor or a lack of willingness on the part of key company decision-makers to finance the plan once strategies have been identified.

If disaster recovery planning is neither as simplistic nor baffling as vendors of plan products and services suggest, it is a complicated task in a field couched in its own terminology and jargon. However, it is the contention of this book, based upon the experience of the author and of many other contingency planning coordinators in businesses throughout the United States, that it is a task that can be managed effectively using basic project management tools and system development concepts.

This book provides a model for managing the disaster recovery effort as a *project*. Using this book and any off-the-shelf project management software package (or even manual project management methods), the reader will have the necessary tools to develop a testable disaster recovery plan. The only boiler-plate offered by this book consists of customizable forms and checklists covering common or generic disaster recovery tasks. The forms used in this book have also been supplied on diskette in a common word processing format so that the reader can modify the forms and checklists to incorporate company-specific requirements or to comply with company documentation standards.

The third purpose of this book is to inspire confidence. Those who are given the task of developing a plan for the survival of the company in the event of disaster typically confront a myriad of dilemmas ranging from a basic fear of the enormity of the task, to difficulties in surmounting the indifference or reluctance of senior management to support the plan. In the face of this adversity, it is not surprising that so many novice planners fail to complete their task or flock to any pundit who promises to validate their efforts with an expensive "certification" program. As a group, the self-confidence of disaster recovery planners is quite low.

This book, recognizes that no expert can honestly promise to deliver a perfect disaster recovery plan. Even those who have successfully managed recovery efforts are quick to point out that their recovery strategies failed in as many parts as they succeeded. Almost invariably, they attribute successful recovery to luck, hard work, on-the-spot ingenuity and innovation, and God. While all believe that outcomes would have been very different had no contingency planning been undertaken, to a person they concede that their plans were—and could only be—imperfect.

Many novice planners voice concern that their plans are somehow inferior because they are "home-grown" rather than "professional." Scanning the marketing brochures of many disaster recovery consultancies, including several Big Six accounting/audit firms, the origin of this feeling is clear: "Joseph P. Consultant has developed more than 100 plans for some of the foremost corporations in the world." What the literature fails to say is that Joseph P. Consultant's plans have been unique for each client, if he has done his job properly. In fact, the best plans are by their nature "home-grown"—customized to meet the needs of the subject company. Regardless of the number of plans that one has developed, each represents a different set of specific recovery objectives and a different set of requirements.

Professionalism is what the planner brings to the plan in the form of work habits, tenacity, and a desire to develop a truly effective recovery capability. It is not measured in numbers of documents developed, but in the quality of a single plan when scrupulously examined and tested.

If every plan is so different, how can a book be written on the subject? The answer is simple. There are generic, cross-cutting principles that apply in disaster recovery planning with which anyone who undertakes planning needs to be familiar. There is also a widely dispersed network of vendors of disaster recovery products and services that may be harnessed to the strategies developed by the planner if the network is known to the planner. Finally, there is instructional merit in depicting an approach to planning that can serve as a guide for the novice, provided such an approach is grounded on concepts and practices with which the reader is already familiar.

THE SCOPE OF THIS BOOK

The objectives of any disaster recovery planning project are two-fold:

1. To develop and document strategies for recovering key systems, networks, and functions of the company that can be tested and modified on an ongoing basis.

2. To alert company personnel and management to the possibility of disaster, to cause them to think about the unthinkable, in order to make them more aware of the disaster potentials that are within their ability to minimize or eliminate.

This book will provide a model for realizing these objectives. The forthcoming chapters treat the development and documentation of disaster recovery strategies from a project management perspective.

What this means is that planning tools with which most business managers and computer professionals are already familiar will be applied to a particular type of project—disaster recovery plan development. This project, like its more traditional business and systems development counterparts, is comprised of objective-based tasks, milestones, testing, and validation components. The project utilizes resources and personnel that must be managed effectively through schedule, budget, and audit controls.

The scope of the project, and thus of this treatment, is company-wide disaster recovery. Therefore, words like "systems" and "networks" must be interpreted in their fullest sense. A system is more than hardware running software. A complete picture of a system includes all sources of input and all types of output. For a computer system to function, certain environmental requirements must be met, such as physical facilities, power, air conditioning, electronic communications, water, and so forth. Also, there must be personnel to operate, manage, and use the computer programs and hardware. The requirements of these personnel must also be considered in a full system description. Moreover, there may be supply and material requirements for outputs—stocks of preprinted forms, and so forth. Thus, systems recovery involves much more than the restoration of a computer platform. It means the recovery of business functions.

The same may be said of networks. A network is more than modems or controllers connected by filaments of wire, coaxial cable, fiber optic cable, and a host of public and private switches. The latter are the mechanical and electronic components of a much more complex collection of human and material resources that work together to make the network meaningful. The entire network, not only the hardware network, must be considered in disaster recovery planning.

Whether a company utilizes a mainframe or a personal computer, a modem or a programmable PBX or a satellite link to provide key business functions, undertaking a disaster recovery planning project is merited. The techniques depicted here will aid in that project.

This book and its supplements will provide insights into the many facets of contingency planning, from project initiation to training, testing, and validation. Not every topic covered will be relevant to every reader, but principles and techniques illustrated here can be adapted and used in practically any company.

MODULAR DESIGN AND CONTENTS

This book is designed as a practical reference for hands-on use by disaster recovery planners. The work is organized into chapters with subsections. Reading the entire book is recommended, but not necessary. Instead, a modular approach allows the reader to pick and choose only those chapters and sections that are of particular interest or relevance to his or her planning needs.

In a sense, the book mimics the disaster recovery planning enterprise. While informational sections are dispersed throughout, hands-on activities that are typically a part of the plan development process are ordered chronologically throughout the book. Following is a brief description of the chapters and their major subtopics.

Chapter 1

This chapter provides a context for disaster recovery planning by examining the rationale for undertaking plans and surveying their benefits and limitations.

Chapter 2

The second chapter sets forth a generic model for a disaster recovery planning project. It identifies and explains the activities comprising the project and elucidates similarities between the planning project and other types of systems development projects.

Chapter 3

The preliminary steps of disaster recovery planning are discussed in this chapter. These steps include research and analysis required to prepare a presentation for senior management on the need to undertake a formal planning project. Also covered are considerations for staffing the project, steps for initiating formal activities, requirements for coordinating and controlling work, and typical project budget requirements.

Chapter 4

Data collection is a key activity in disaster recovery planning. This chapter provides a sample methodology for collecting relevant information on business functions and their systems and network and records requirements.

Chapter 5

Chapter Five examines the component tasks of exposure and risk analysis. Guidance is offered for use in conducting threat assessments, defining exposures, and calculating maximum acceptable downtime. Risk analysis methods are surveyed and insurance is considered as a disaster impact mitigator. Additional advice is provided on presenting the findings of risk analysis to senior management.

Chapter 6

An important component in effective disaster recovery planning is identifying avoidable disasters and implementing prevention capabilities to minimize the likelihood of such disasters. This chapter suffests an approach to identifying and implementing disaster avoidance requirements.

Chapter 7

Routine backup and off-site storage of mission-critical data is arguably the most important ingredient of effective disaster recovery planning. In this chapter, readers will be introduced to a method for determining the importance of data and records and will be provided with a detailed discussion of backup and storage strategies.

Chapter 8

Systems recovery is the traditional focus of disaster recovery planning. However, as distributed computing takes hold in corporations worldwide, mainframe-focused systems recovery strategies face new challenges. This chapter offers guidance for planning based on the development of a minimum equipment configuration. Recovery strategy alternatives are set forth and evaluative criteria are offered.

Chapter 9

This chapter examines network recovery planning. This aspect of planning is often ignored by traditional disaster recovery, except insofar as communications requirements exist for system restoration. In fact, critical business functions are increasingly served by networks themselves and interruptions in network operations can constitute a disaster for many organizations. This chapter guides readers through a procedure for defining minimum service requirements and for evaluating options available for network restoration.

Chapter 10

In Chapter Ten, a strategy for end user recovery is examined. End users, like networks, are often ignored in traditional disaster recovery planning thought. Yet, it is the end user work area where a disaster is most likely to occur. Without the restoration of end users, system and network recovery efforts are pointless. The concept of a backroom operations center is introduced and readers will be introduced to the logistics and planning requirements for this approach.

Chapter 11

Once disaster avoidance system requirements have been identified, backup and off-site storage plans formulated, and strategies for system, network and end user recovery articulated, it may be necessary to present these recommendations to senior management for approval and implementation. This chapter provides guidance for making such a presentation and for formulating a schedule of implementation and maintenance tasks and costs.

Chapter 12

This chapter examines techniques for writing an effective disaster recovery plan document. Various plan structures are considered and examples are provided. Additionally, guidance is provided on the formulation of an Emergency Decision-making Flowchart and the development of Notification Trees.

Chapter 13

Training is an integral component of disaster preparedness. This chapter sets forth a training development methodology based upon principles of adult education and instructional systems design. Also covered are awareness programs that can enlist all corporate personnel in the disaster prevention effort and opportunities for continuing education in the field of disaster recovery.

Chapter 14

Testing is a critical component of maintaining an effective disaster recovery capability. This chapter examines the goals of testing and types of tests and offers guidance on conducting tests and obtaining useful results. Also described is a method for managing changes to the disaster recovery plan itself.

Use of Illustrations

Illustrations in this book consist of flowcharts and other diagrams depicting concepts described in the text. Also, sample forms and checklists are provided for the reader's use in performing planning tasks. These forms have been provided on diskette and may be modified using several popular word processing and spreadsheet software products currently available.

Invitation for Comments and Criticism

This book is the product of the shared experiences of many disaster recovery planners as well as vendors of plan products and services. It is also a distillation of the author's own experience both as a contingency planner and as the co-founder and chairperson of a professional organization for disaster recovery planners in Florida. As such, the views and the approach represented in this book are the synthesis of many into one.

Along the way, decisions needed to be made about what information was germane to the majority of businesses and what was not. Every disaster recovery planning project has its own unique problems and solutions, however. To the extent possible, this book provides a flexible context for treating company-specific needs while providing a solid foundation for meeting cross-cutting issues and obstacles head on.

Still, there may be omissions or biases expressed here that require correction or refinement. There are also technological developments that may require additions and amendments to this text over time. For this reason, readers are encouraged to express their comments and criticisms of the book, preferably in writing, to the author.

Introduction

DEFINITION OF A DISASTER

For the purposes of this book, a disaster is defined as an interruption of mission-critical information services for an unacceptable period of time. This definition reflects the nature of disaster and avoids the problems that frequently arise by applying categorical adjectives to a disaster.

The nature of disaster is relative and contextual. What constitutes a disaster for company A may not be a disaster for company B. Whether the result is a hard disk failure on a PC managing the accounts receivable function of a small business or the loss of data processing facilities of a major company to a fire, hurricane, or earthquake, it is a disaster if the event causes an interruption of mission-critical information services to a firm.

Efforts have often been made in disaster recovery literature to categorize disasters by their scope or origin. One reads of facility or localized disasters, regional disasters, natural or man-made disasters. These terms, while descriptive, do not provide insight into what qualifies an event as a disaster. Only if an event interrupts the means by which critical corporate functions are accomplished can it be termed a disaster. Given the fact that most mission-critical functions in the modern business world are accomplished using automated information systems and networks, a disastrous event is typically characterized by an interruption of the means by which business data is produced, processed, and/or distributed.

Disasters are also defined in relation to time. Time is important from the standpoint of when an interruption occurs and how long the interruption lasts. An interruption of telephone service to a telemarketing company during a peak sales period is probably more disastrous than the same interruption during nonbusiness hours. A data processing interruption that occurs during year-end processing may create a greater problem than one that occurs during a period of the month when processing demands are less. In both cases, the timing of the interruption plays a role in defining it either as a disaster or merely an inconvenience.

Similarly, the duration of an interruption often defines its characterization as a disaster. A power outage of several hours may strain the data processing capability of a company without a backup generator, but with hard work and some expense, the company may be able to absorb this event without much harm. However, a power outage of many days, owing to a cut cable or a cataclysmic hurricane, may not be so readily tolerated and may have disastrous consequences.

What constitutes an "unacceptable period of time" for the interruption of information services is a relative thing. It is based on a simple cost-tolerance effect. Over time, the costs to the company of an outage increase and the company's tolerance to the interruption decreases. This fact has been demonstrated statistically by numerous academic and industry-sponsored studies.

The natural question that readers may ask based on the previous definition is, "How do I determine what constitutes a disaster for my company?" There are many factors to consider and many ways to interpret and present the data, but this is one calculation that becomes very specific to a company. This book will provide examples and alternative approaches to assessing the prospective impact of an interruption but, in the final analysis, the reader will need to arrive at his or her own conclusion about the tolerance of the company to an interruption in information services.

RATIONALE FOR DISASTER RECOVERY PLANNING

Given this brief discussion of disaster, it may occur to some readers that disaster recovery planning is a misnomer. An interruption from which a company can recover, via some planned strategy, is not properly termed a disaster. A disaster recovery plan provides a capability to the company to absorb the impact of an event and to prevent business failure. For this reason, new jargon has been introduced to describe the disaster recovery plan as a business continuation plan or simply as a contingency plan.

However, disaster recovery planning is still the most recognized term for this business planning enterprise; the loss of life and resources that often accompany catastrophic events may indeed earn these events their name—disasters. Certainly, without disaster recovery planning, an interruption in mission-critical business functions stands a much greater chance of translating into a disaster for the business.

Conversely, there is substantial evidence to support a case that those companies which plan for the possibility of a disaster stand a far better chance of recovering from a catastrophic event than those that do not. This is the ultimate rationale for disaster recovery planning.

In case study after case study, companies that have survived disasters attribute their success to the implementation of pre-planned strategies by personnel who have been trained and rehearsed in their recovery roles. (Even in these cases, however, there are common references to on-the-spot innovation, luck, and God to help explain recovery. Those who have experienced disasters are almost invariably humbled by the event, rather than viewing their recovery as a testimony to their planning skills and foresight.)

The disaster recovery plan itself provides certain logistical supports for recovery that would be difficult to assemble *on-the-fly* in the wake of disaster. These supports may include prearrangements for an alternative data processing site, for on-demand rerouting of data communications traffic, for power, for alternative work areas, and for recovery of system databases from off-site backups. Moreover, plan training and testing forces those who will play a role in plan implementation to confront the dark side—the possibility of disaster—and to be better prepared to respond rationally to the great irrationality and chaos that often accompany crisis.

Beyond these substantial contributions, effective disaster recovery planning—which includes the identification and implementation of disaster avoidance capabilities and the articulation of disaster awareness training—may help to prevent avoidable disasters. Thus, the effectiveness of disaster recovery planning may also be measured in *non-events*—disaster potentials that have been minimized or eliminated.

This rationale for disaster recovery planning falls short of a blanket guarantee that a company will always recover from any disaster. Still, it provides some measure of recovery assurance that is missing in the absence of such planning. In short, disaster recovery planning constitutes the only strategy at our disposal for coping with the unpredictable occurrence and consequences of disaster.

Still, business managers (and many information systems professionals) may view disaster recovery planning as an unjustified expense based on the previous assessment. Working against expenditures for planning may be a statistic, long advanced by insurance companies, that places the possibility of a disaster at one percent per annum. To those preferring to play the percentages, investing resources in disaster recovery planning may be difficult to justify.

Thus, in addition to common sense, there often need to be other rationales for undertaking this effort. These may include the requirements of user contracts, auditors, or law.

User Contracts

In many companies, data processing and other information services are provided to end-user departments on the basis of a contract called a *service level agreement*. Service level agreements establish the ground rules for interaction between company data processing and the user department. Agreements often include descriptions of procedures users will observe when requesting services from data processing and also of *charge-back* or other schemes that will be employed to repay data processing for its work.

Such agreements are considered an important component of company cultures that seek entrepreneurship and innovation from their work units. Data processing, like any other work unit, is expected to provide the best possible service at a cost that is competitive with outside vendors of information processing services. Service level agreements, though not legal documents, formalize this value and promote it.

Many service level agreements also provide a guarantee of service availability. This may mean promising the user that systems will be available for use 80 percent of the time or it may be more simply expressed as a goal of zero downtime.

Such promises help to justify the shared expenditure of budgeted resources for company data processing services and cannot be made in good faith without a provision for disaster recovery planning. Some disaster recovery planners have used the rubric of the service level agreement not only to justify disaster recovery planning, but also to charge back the costs for such planning to user departments. Sometimes this is done so that the departments with greatest system usage absorb more of the expense than those that use services minimally.

The shifting focus of disaster recovery planning from information systems to the more comprehensive objective of business function recovery may make departmental chargeback more and more viable. This, in turn, may diffuse management reluctance to shoulder the expense of planning. The reasons for this will become clearer with a discussion of the collateral benefits of disaster recovery planning for departmental efficiency; suffice it to say that if disaster recovery planning is not perceived as a luxury expense for a single department, but instead as a shared requirement of all business units, political—as well as rational—arguments will favor management support.

Audit Requirements

For companies that do not use service level agreements or chargeback schemes, the previous rationalization of disaster recovery planning expenses may be insufficient to justify management expenditures. In these cases, a second rationale may be forthcoming from company auditors.

In many companies internal auditors serve as "eyes and ears" for senior management, alerting them to issues of legal liability, financial exposure, and business function efficiency. Some companies contract this work out to external auditors whose periodic visits provide third-party insights into what is right and wrong with company operations.

Auditors who wish to examine the operational characteristics of a company are increasingly called upon to scrutinize the information systems and networks which provide critical company functions. Here, they look for evidence of data integrity—that data is being entered, manipulated, and output without error. Of particular interest to auditors is system and network security, a field closely akin to disaster recovery.

Auditors perceive the value of information in the modern business and vulnerability of the company should information become suspect or corrupt. The auditor requires proof of the information system manager's attention to security and checks all security measures for fail-safe operation.

Recently, the auditor's preoccupation with data integrity has begun to include an interest in contingency plans. While a substantial threat to data integrity is found in the disgruntled employee or computer hacker, auditors recognize other threats as well. Flooding is responsible, statistically, for most computer downtime according to the National Fire Protection Association (NFPA), followed by fire and other disaster causes. Auditors are increasingly interested in what measures a company has taken to safeguard its information resources from these disaster potentials, as well as from traditional security risks.

Most Big Six auditors will cite the lack of a disaster recovery plan at a client site. Some will go so far as to cite the absence of records within the plan (in those client companies that have one) that indicate the plan is regularly tested and maintained. A few will ask to participate in plan testing, playing the role of observers, so that plan solvency can be evaluated.

The absence of a disaster recovery plan will generally be noted in the audit comments with the auditor's specific recommendations regarding the need to address this deficiency. If management is swayed by audit recommendations, this comment may provide a sufficient rationale to initiate and support planning efforts.

Legal Requirements

Some auditors will point to the absence of a disaster recovery plan not only because of the need that they perceive based on threats to the information asset, but also in response to legal mandates. In some industries, particularly in financial institutions that participate in the Federal Reserve System or federally-backed deposit insurance programs, disaster recovery planning is required by law and company compliance is rigidly enforced. For other industries, the legal mandate is less clear and rarely enforced unless as part of a larger issue, such as illegal foreign trade.

In financial institutions, failure to comply with the legal mandate—embodied in Office of the Comptroller of the Currency banking circulars—may expose companies to sanctions and fines, and bank managers to personal liability. Federal bank examiners have been paying close attention to bank adherence to these requirements in the wake of a computer system failure at the Bank of New York in 1985 that temporarily destabilized government securities.

Government contractors are similarly required to take steps to safeguard their systems by Office of Management and Budget (OMB) Circular A–71. Particularly relevant to Department of Defense contractors, the rationale behind the circular is an acknowledgment that most of the design and production of military equipment and other goods is accomplished using computers; thus, systems and data need to be protected in the interests of national security.

Following passage of the Computer Security Act of 1987, the provisions of OMB Circular A–71 were given more teeth. In compliance with the Act, OMB issued Bulletin 88–16 requiring that security plans (including "emergency, backup and contingency plans") be prepared by all government agencies and their contractors for submission and review by the National Bureau of Standards (now the National Institute for Standards and Technology—NIST) and the National Security Agency (NSA). NIST/NSA review teams will be classifying government systems based on the sensitivity of the information they handle and evaluating plans that have been formulated to provide security and recoverability for those systems. The Computer Security Act further provides for training assistance to agencies to assist them in developing plans where none currently exist. The result of this effort is expected to be a uniform planning standard for government systems with which agencies and their contractors will need to comply.

For other industries, the legal mandate for disaster recovery planning is somewhat less compelling. The Federal Foreign Corrupt Practices Act of 1977 requires all companies to take measures to safeguard and guarantee the integrity of accounting and ledger information stored and processed on data processing systems. The act provides for fines of up to $10,000 and five years imprisonment for individual business managers and corporate penalties of more than $1 million for companies that fail to comply with the act. However, enforcement of these provisions is expected to be limited to cases involving businesses with questionable overseas sales.

All companies in the United States are also subject to an assortment of Internal Revenue Service (IRS) requirements regarding the preservation of machine-readable financial records. Numerous IRS procedures and rulings require businesses to retain and safeguard records that may become material in an IRS audit or in the administration of IRS law.[1] Penalties for noncompliance with these procedures and rulings are typically directed at company officers.

In addition to the previous regulations, guidelines and regulations of interest to auditors have been articulated by the Occupational Safety and Health Administration (OSHA), the Environmental Protection Agency (EPA), and the Small Business Administration (SBA). Over the past five years, there has also been activity in state legislatures in the area of computer security and contingency planning. Auditors should be versed in these and all requirements that affect their companies or clients.

Auditors will typically bring to management's attention any instance of illegality or regulatory noncompliance that they encounter. This may include the failure of a company to comply with federal and state laws on disaster recovery and computer security. Given the exposure that such noncompliance may create, and the increasing tendency of regulatory agencies to hold management personally accountable and legally liable, this can provide a compelling rationale for disaster recovery planning.

ADDITIONAL BENEFITS

The rationale for disaster recovery planning may thus include contractual ethics, audit recommendations, and legal mandates. These, however, are secondary to the main rationale grounded in common sense: Companies with tested disaster recovery plans tend to survive disasters, while those without such plans do not. To many planners, this rationale is so self-evident that they cannot envision management reluctance to sponsor such a worthwhile project.

Still, convincing senior management to underwrite the costs of the disaster recovery planning project may require that other advantageous by-products

---

[1]The most important IRS mandates to date include Procedure 64–12, which covers the requirements for secure storage of recorded or reconstructible data, and Ruling 71–20, which extends these requirements to machine-readable records.

of the enterprise be identified. Following are some additional benefits that may accrue to disaster recovery planning:

1. **Business Insurance Cost Reductions**—Effective disaster recovery planning can reduce business insurance costs in two ways. First, the formal risk analysis performed in disaster recovery planning can fine tune the coverages purchased from a business interruption insurer. Often, in the absence of formal risk assessment, the identification of insurable risks is conducted in gross terms, resulting in the acquisition of inappropriate or overly expensive coverages. Thus, disaster recovery planning can provide data to better target business insurance and thereby reduce aggregate insurance costs.

    Another way that disaster recovery planning can aid in business insurance cost reduction is through its emphasis on disaster prevention. By identifying risks and installing detection and suppression systems (Halon or sprinkler systems, water detection systems, etc.) to handle them, disaster recovery planning can provide a demonstrable disaster prevention capability. This capability, in turn, may provide a rationale for convincing insurers to reduce their insurance prices. Often, the existence of a tested disaster recovery plan will achieve the same results.

2. **Business Uses of Disaster Recovery Facilities**—In some cases, facilities and equipment obtained to facilitate disaster recovery may be used for training and other purposes. Where possible, these normal business uses for contingency capabilities should be stressed.

    Also, certain disaster avoidance systems, such as power conditioners or uninterruptible power supplies, provide services that will improve the performance of protected systems during normal use. If, for example, short-term outages due to brownouts or thunderstorms are a problem, these disaster-avoidance systems will help to prevent them and improve overall system uptime statistics during day-to-day processing.

3. **Promotion of Corporate Culture Goals**—A company that undertakes disaster recovery planning can leverage this activity in a number of meaningful ways. Marketing units may be able to reference their company's attention to contingency planning as a demonstration of a commitment to product or service availability. Company planners may speak at conventions or in the media about the merits of disaster recovery planning, promoting the company as a leader in the field. Part of the disaster recovery plan may include the establishment of a community emergency shelter on company premises, an excellent platform for articulating the value that the company places on good corporate citizenship. While some may regard these "uses" of company disaster recovery planning as cynical, they are also a valid and beneficial means for disseminating information about disaster recovery planning and influencing other companies to follow suit.

Other benefits of disaster recovery planning may include improved records retention policies, improved documentation on company systems and networks, and a host of other by-products that can improve business performance or enhance management understanding of business functions. Planners should be attentive to such collateral advantages and use them as additional justifications for management sponsorship.

LIMITATIONS OF DISASTER RECOVERY PLANNING

Disaster recovery plans are not perfect. Even with regular testing, plans may contain flaws that reveal themselves only in a disaster. At best, a disaster recovery plan provides most of the required logistics for recovering from an interruption. Hopefully, regular testing and training equip plan participants with a mindset that will aid them in dealing with the mind-numbing reality of disaster. However, emergency action plans and recovery procedures are guides and not models, and are subject to change based upon the actual consequences of a disaster.

PRAGMATIC ORIENTATION OF EFFECTIVE PLANNING

In accordance with the limitations previously described, the orientation of disaster recovery planning must be pragmatic. This pragmatism reveals itself in several ways.

Plan Scenario and Objectives

Developing a disaster recovery plan requires that threats be identified and strategies be developed to respond to them. First-time planners often cite difficulties with this central point. Lacking a pragmatic orientation, planners become mired down in planning for every conceivable disaster scenario. The result is often a failed planning effort with the size and scope of the plan exceeding all realistic dimensions.

What is a pragmatic orientation? The recognition that one cannot plan for every contingency leads to the conclusion that one should plan for the worst-case scenario. A plan oriented toward worst-case provides procedures for dealing with less devastating events as well as *smoke-and-rubble* catastrophes.

Such a plan would provide procedures to meet a generic set of plan objectives, summarized in Figure 1.1. Procedures would be developed to evacuate personnel in unsafe situations, to recover user work areas, to recover systems and networks, to back up and safeguard vital records, and to salvage damaged facilities or locate and equip new ones. These procedures could be implemented in whole or in part to cope with a range of disaster scenarios. For example, if a fire damaged the company data center, but left user areas and other voice and data networks unscathed, the disaster recovery plan could be implemented in part to recover critical system operations. Similarly, if a local telephone company serving office were struck by a disaster, but company facilities were intact, only network recovery components of the plan would be activated.

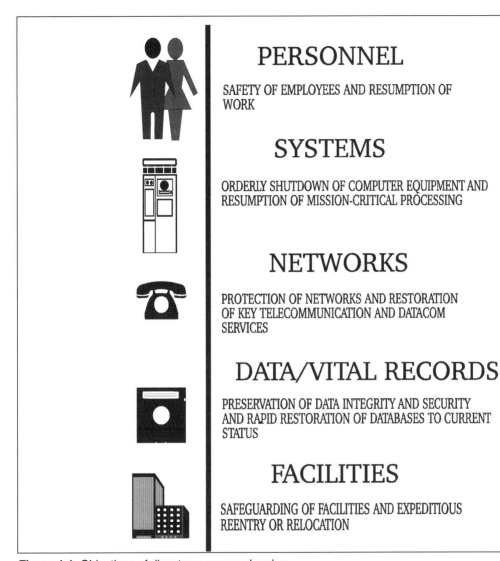

PERSONNEL

SAFETY OF EMPLOYEES AND RESUMPTION OF WORK

SYSTEMS

ORDERLY SHUTDOWN OF COMPUTER EQUIPMENT AND RESUMPTION OF MISSION-CRITICAL PROCESSING

NETWORKS

PROTECTION OF NETWORKS AND RESTORATION OF KEY TELECOMMUNICATION AND DATACOM SERVICES

DATA/VITAL RECORDS

PRESERVATION OF DATA INTEGRITY AND SECURITY AND RAPID RESTORATION OF DATABASES TO CURRENT STATUS

FACILITIES

SAFEGUARDING OF FACILITIES AND EXPEDITIOUS REENTRY OR RELOCATION

Figure 1.1 Objectives of disaster recovery planning.

A plan designed in this manner is manageable and effective. Plus, its sectional design is efficient in other ways. Recovery personnel can be assigned plan sections to learn, test, and maintain according to the role they are expected to play in a disaster. Moreover, when an interruption occurs, sections of the plan may be distributed to recovery team leaders rather than reproducing many copies of the whole document.

Risk Analysis

A pragmatic approach is also effective in risk analysis, perhaps the most misunderstood aspect of disaster recovery planning. In some quantitative approaches to risk analysis, an unrealistic effort is made to assign exact dollar values to various downtime scenarios or to calculate to the penny the cost to the company of a minute of system or network interruption. It is perhaps believed that by giving the illusion of scientific accuracy to an assessment of vulnerability, the disaster recovery requirement can be made more compelling for senior management. While this argument may have merit, its corollary does not: It is not possible to quantify in advance the financial impact of a disaster nor to provide a one-to-one correspondence between dollars spent on recovery planning and dollars saved in the event of disaster as the result of such planning.

The pragmatic approach is to generate—based on a creative compilation of downtime cost estimates taken from user departments, company sales figures, and other sources—a ball-park estimate of loss potentials. Typically, these figures provide a compelling cost justification to a receptive management even without the guise of mathematical accuracy. Moreover, loss potential estimates provide a general guide for determining how long a company can sustain itself deprived of critical systems and networks. This number, in turn, may be used to establish a maximum allowable downtime for the company and provide a key criteria for the development of recovery strategies.

Emergency Decision-making

Another contribution of healthy pragmatism in disaster recovery planning is the view of emergency decision-making and other aspects of plan implementation that should permeate both the plan document and training programs that are used to rehearse and prepare plan participants. Case studies repeatedly show that, because plan solvency is limited by so many unpredictable factors, such components as an emergency decision-making flowchart must be taken as guides rather than models or scripts.

Decision-making flowcharts provide a general chronology of likely recovery events and are not intended to be followed slavishly, but to provide a baseline for innovation and flexible response. Similarly, plan procedures provide a solid basis for recovery activity, but recovery teams may need to stray from documented procedures should situations dictate.

Recovery Strategies

Being pragmatic means selecting recovery strategies that will ensure timely and effective restoration of business functions, but will not be more costly than the value of the functions themselves. There is a trade-off between expense and confidence in strategy formulation. Strategies that afford highest levels of confidence in recovery are also among the most expensive. For example, building a second, identical data center at a remote location is a high confidence strategy for system recovery. However, the expense of such a strategy is prohibitive for all but a few companies and organizations; even these few are rarely able to rationalize the expense on the grounds of disaster preparedness alone.

On the other hand, if redundancy is the only effective means for ensuring recovery of a critical function, a pragmatic approach to justifying this expense often includes other uses of the redundancy. For example, if a company uses a piece of computer hardware that is critical for business operation and for which a replacement would be difficult to obtain in an emergency, obtaining a second unit would be necessary. However, this installed spare may be justified not only in terms of its usefulness in an emergency, but also as training apparatus or as a source of spare parts for the primary unit.

Methodology and Project Controls

Development of the disaster recovery plan is a multifaceted project involving the coordination and management of diverse resources and the creation, testing, and maintenance of numerous project deliverables. As previously mentioned, many approaches and planning methodologies have been developed to aid in the development of a plan. Most require of their users that they become specialists in the peculiarities of the methodologist's preferred techniques.

Pragmatism argues against these approaches. Using a planning methodology that requires skills not present in the typical repertoire of business planning techniques creates a learning curve that the planner, and eventually his or her replacement, must surmount. With people changing employment about every three years on average, it makes little sense to complicate the job of disaster recovery planning by using a byzantine methodology. This, in a nutshell, is the pragmatic case for a project management approach.

Pragmatism should also guide such project control features as the electronic format of the plan document and the media on which it is stored. Many disaster recovery planning tools reside on mainframes or utilize less familiar database or integrated software packages to store data. Microcomputers are far more abundant than mainframes and an ASCII text document, accessible with any popular word processor, is more user-friendly than a more complex database management system. Unless there are compelling reasons to the contrary, a diskette-based text file is generally the most agreeable electronic format for the disaster recovery plan.

Pragmatic considerations both simplify and render more accessible the task of disaster recovery planning. To the extent possible, this standard embodies the principle of pragmatism.

2

The Disaster Recovery Planning Project

Up to now, this book has touched on some of the general aspects of disaster recovery planning. At this point, we need to introduce the planning model that will be used to represent the major activities undertaken in a typical project.

MULTIPHASIC APPROACH

From a linear perspective, the disaster recovery project has five phases. Figure 2.1 depicts these phases as project initiation, risk analysis, strategy development, plan implementation, and support and maintenance. This is how the project is often illustrated in vendor literature. It has a beginning, project initiation, consisting of two basic activities: convincing management that disaster recovery planning is needed and assembling a team or hiring a consultant to do the job.

Risk analysis, the next phase, typically translates to developing a list of critical systems and networks, identifying threats to these assets in the form of natural and man-made disasters, associating a dollar value with the interruption of these systems and networks, and developing restoration criteria, consisting of a maximum allowable downtime statistic or some other time-specific value.

The next phase is strategy development. Chief activities of this phase include identifying alternative means for providing necessary system or network services (typically amounting to a survey of disaster recovery products and services) and soliciting custom bids from vendors.

Plan implementation is interpreted as the signing of contracts with selected disaster recovery vendors. It may also include the implementation of programs of regular data backup and off-site storage. Finally, the project concludes with plan support and maintenance. Ideally, this will include the training of those who will participate in the plan should a disaster occur, the periodic review of plan provisions and target systems to account for changes in hardware and software, and periodic plan testing.

This linear model offers a simplistic presentation of major plan development activities. It is sometimes useful as a tool for presenting the plan to management. However, as a guide for those who must undertake the development of disaster recovery plans a relational, rather than a linear model, is needed. In fact, reality is considerably more complicated than a linear diagram would suggest. Borrowing from the toolbox of structured systems analysis, Figure 2.2 provides a better model for the disaster recovery planning enterprise.

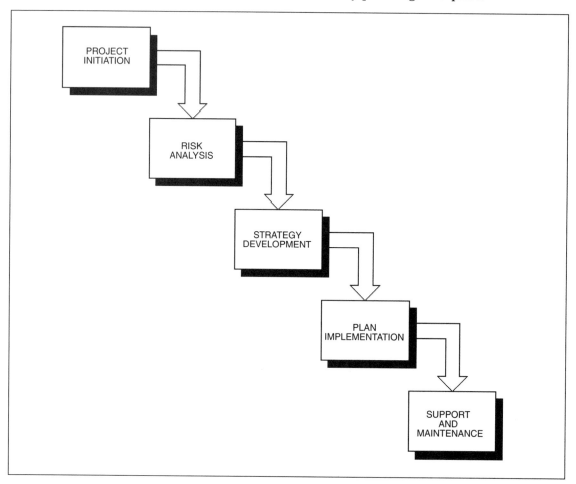

Figure 2.1 Linear model of the disaster recovery planning project.

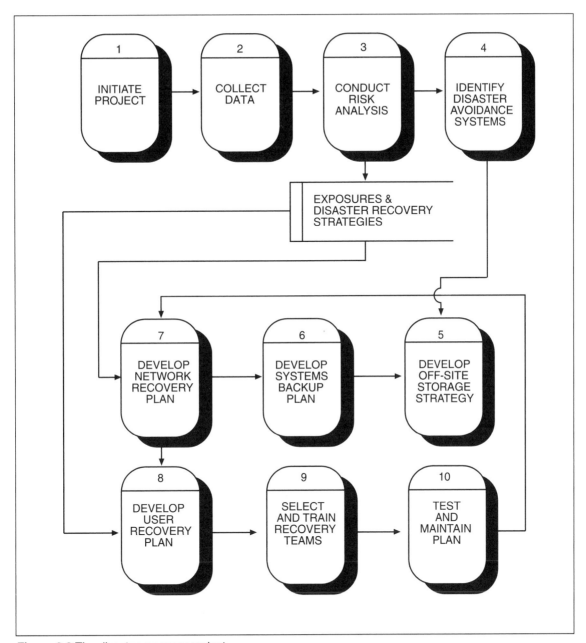

Figure 2.2 The disaster recovery project.

In this model, planning is shown to consist of nine discrete activities. Directional arrows show the interdependencies of these activities and illustrate feedback loops that make disaster recovery planning an iterative process.

As in the linear model, planning begins with project initiation. As we shall see in Chapter 3, this activity consists, in large part, of many subtasks that together comprise an informal risk analysis. The purpose of this informal risk analysis is to communicate to management that there is a need to review disaster preparedness at the company, and not necessarily to solicit blanket approval on the planning project. With management approval, a project team is assembled, an action plan is developed, and coordination and control measures are established. A budget is also developed for the formal risk analysis phase and this, plus the selected project leader or planning coordinator, are presented to management for approval.

Risk analysis is conducted in two parts: data collection (Chapter 4) and exposure and risk assessment (Chapter 5). Data collection activities aim at producing a comprehensive picture of business functions, depicting resource requirements for each function as well as functional interdependencies. Exposure and risk assessment consists of documenting disaster potentials, identifying and estimating costs of work interruptions, and developing a worst-case scenario for presentation to senior management.

Figure 2.2 shows an arrow moving from the "Conduct Risk Analysis" symbol to a symbol, called a data store, labelled "Exposures and Disaster Recovery Strategies." This is a diagrammatic contrivance. A *data store* is a repository for information that may be used in other phases of disaster recovery planning. In reality, the data store may consist of research materials, case studies, user questionnaires, and risk analysis documents that may be referenced as suitable recovery strategies.

Still another activity of risk analysis is to develop a *spectrum of criticality*—that is, a range of values indicating how difficult it would be to make do without a business function over a period of time—and to assign each business function a value on this spectrum. This is among the most difficult tasks in disaster recovery planning since all employees tend to view what they do as critical to the business. In fact, criticality is based on the dollar losses that would accrue to the company if a subject function were not performed for a given period of time. If a function could be interrupted for 48 hours without causing the company to fold, it is probably not critical. It may not, therefore, have the same restoration priority as another function that would cause the business to fail if it were interrupted for a comparatively brief space of time.

The results of risk analysis are often presented to senior management to solicit approval to proceed. If management perceives the need for disaster recovery planning following an honest presentation of exposure data, and if it is otherwise predisposed to allocate resources to developing a disaster recovery capability, work may then proceed on formulating strategies for backup and storage, system recovery, network recovery, and user recovery.

Having formally analyzed company exposures, it is important to determine the extent to which disaster potentials can be minimized or eliminated through the acquisition of disaster avoidance systems. In some cases, this activity, which is covered in detail in Module Five, may be undertaken before risk analysis results are presented to senior management. In this way, exposures may be used to rationalize both a disaster prevention capability and the disaster recovery planning project. Pragmatism dictates that monies should be spent first on preventing avoidable disasters, then on planning for the recovery of the company from disasters that cannot be avoided.

Disaster avoidance begins with facility design considerations, over which the disaster recovery planner may have no control. Included in a disaster avoidance capability are environmental contamination controls; power and other support system continuity measures; detection, alarm, and suppression subsystems for fire, flooding, and other disaster potentials; and equipment backups and spare part logistics.

Having identified (and acquired) disaster prevention capabilities, the next disaster recovery planning activities involve the formulation of an off-site storage plan, system recovery plan, network recovery plan, and user recovery plan. Bidirectional arrows in the flow diagram (Figure 2.2) suggest that these activities are interrelated. For example, the network strategy must consider the locations of user work groups and systems if the latter have moved to alternate facilities following a disaster.

Note also that a directional arrow connects the data store symbol to the recovery strategy activities. Data about systems, networks, and business functions, collected in risk analysis, is referenced by planners as they formulate recovery strategies for these resources. How recovery strategy development is accomplished is covered in Chapters 7 through 10.

Once strategies have been formulated, they are documented. Costs for obtaining necessary commercial products and services for use in each strategy are calculated for later presentation to management. Before the presentation, however, it may be valuable to examine current business insurance coverages. Chapter 11 provides a brief discussion of business insurance coverages and identifies some specific ways that disaster recovery planning can reduce insurance costs. These cost-savings should be documented and presented to senior management together with strategies and their costs.

Chapter 11 offers guidance on this critical presentation. It is at this point that management proves its commitment to disaster recovery planning by approving the acquisition of planned services and by authorizing expenditures for plan implementation, training, and ongoing plan testing and maintenance programs.

Plan implementation (see Chapter 12) involves the creation of the procedural document that embodies the recovery strategies previously developed. It also entails the development of an emergency decision-making flowchart that attempts to depict the chronology of tasks that will be undertaken in an actual disaster. Finally, implementation includes the identification and selection of recovery teams, personnel who will play a role in activating the disaster recovery plan in the event of a catastrophe.

Note that the implementation tasks, including the insurance review, previously described do not have a corresponding activity symbol in Figure 2.2. This is because practically every company performs these tasks in different ways. In many instances, procedure and emergency decision-making flowchart development, and even team selection, occur during strategy development. Also, the complexity of business functions varies from company to company, so the strategies for recovery may be relatively straightforward or quite technically complex. This, in turn, may mitigate or necessitate the need for a separate plan implementation activity.

The final activity symbols in the flow diagram involve team training (discussed in Chapter 13) and plan testing and maintenance (Chapter 14). For too many companies, disaster recovery planning ends when a paper document has been developed and is placed on the shelf for auditor review. To translate plans into a capability for recovery, plan participants must be trained in their roles and rehearsed in the procedures that they may be called upon to perform through periodic testing. The results of tests and ongoing information about system, network, and functional changes within the company must be analyzed for their impact on current recovery strategies and integrated into the plan to keep it up-to-date.

SIMILARITIES TO OTHER BUSINESS PLANNING PROJECTS

The multiphasic model of the disaster recovery planning project just described is not much different from other business planning functions. For example, when a manufacturer plans to introduce a new line of products or a retail mortgage banking company plans to enter into a wholesale operation, substantial research and analysis is conducted to set parameters on the venture. Then, strategies for integrating the new business functions into the organizational structure are developed. Finally, a cadre of employees are trained to support the function and market testing is conducted to refine and repackage the product for best performance.

Disaster recovery may be regarded as a new business function. Of course, the product produced by the disaster recovery project is mainly for internal consumption, likening it to a business system development project. In fact, it is this comparison to system development that may shed the greatest light on disaster recovery planning. In developing a new automated system, first management must be convinced that the new system will contribute something to the company—greater efficiency, lower operating costs, and so on—that would not be possible without the system. New systems, like disaster recovery planning, must be justified in *business terms* and not merely on the basis of technological trends and directions.

Projects that receive management support proceed with a careful analysis phase, in which business needs are scrutinized, and data inputs, manipulations, and outputs are carefully documented. Moreover, the systems analyst will review the current hardware and software platform to determine whether resources are adequate to support the operation of the new system.

In a sense, these procedures constitute a risk analysis. Systems analysts and systems programmers are seeking to ensure that the proposed development will not compromise the company's investment in technology and that new applications will perform required functions for a sufficient period of time to justify their developmental expense. These determinations may be made somewhat more readily than the cost-advantages and risk-reduction benefits of disaster recovery planning.

With analysis complete, a design phase is initiated in system development, possibly resulting in a model or prototype of the new system. This prototype may be loosely equated to the strategies that are developed in disaster recovery planning. The prototype provides a vehicle for communicating operational characteristics about the new system to management and to key end users, both to solicit approval and to obtain comments and criticisms.

When the design is approved, resources are committed to coding the actual system. Similarly, in disaster recovery planning, strategies are documented and commercial products and services are acquired.

With the system completed, it must be validated. Extensive testing is performed until a confidence level has been reached that the system is ready for release. In some system development projects, a group of end users are trained in system operation so that they can aid in testing, just as plan participants are trained so they can participate in disaster recovery tests. Test results are reviewed and integrated into the system to resolve errors or enhance capabilities.

Finally, over time, support and maintenance tasks are performed, both in the case of new application systems and disaster recovery plans, that aim at keeping the capability current and relevant to the business. Feedback channels are established to ensure that system managers are advised of company plans that will impact on their systems. These changes are evaluated and systems may be modified to accommodate them.

This parallel between the systems development life cycle and the disaster recovery planning life cycle may be of use to information systems managers and others who are called upon to develop a disaster recovery plan. The methodology employed in systems development is readily translated for use in developing a disaster recovery plan.

INTEGRATION WITH OTHER PLANNING AND DEVELOPMENT PROJECTS

This book deals with the initial development of a disaster recovery plan. As should be apparent by now, disaster recovery planning is an ongoing effort. As new systems and business functions are introduced at a company, the shape of the disaster recovery plan may need to change. The change may be evolutionary or radical, depending in large part on how successful the disaster recovery coordinator has been in communicating disaster recovery rationales and requirements to business and information systems managers.

For example, information systems managers need to be made aware of the potential impact of new systems on the disaster recovery plan and to consider this impact as they evaluate the costs for implementing new systems. In most cases, this impact analysis will amount to a line item in the proposed budget for the new system. Rarely will disaster recovery costs constitute a make-or-break factor in determining whether a new systems project is worth undertaking.

However, it is vital that the disaster recovery planner be involved in all new system (or network) development efforts so that costs may be anticipated. One can envision a situation in which a company decides to rearchitect its systems along client-server or other decentralized models and makes plans to convert all existing applications to operate under a decentralized database management system (DBMS). In such a case (and even in the case of some less-revolutionary system changes), the impact on the current disaster recovery plan can be devastating. All strategies may need to be modified or entirely new strategies developed to accommodate the requirements of the new systems.

In fact, if the disaster recovery planner can become part of the team for the conversion or new system development effort, the disaster recovery capability of the company need not be adversely affected at all. The planner can work beside systems analysts and technical architects and make use of the same specifications to develop new recovery strategies.

The planner may also be able to contribute insights gleaned from the discipline of disaster recovery planning that will be useful to systems development personnel and at the same time simplify the task of strategizing for system recoverablility. For example, by suggesting methods to enhance the performance of the system—such as fault prediction/fault tolerance software, modular software design, and even alternative hardware selections—the planner can also provide for solutions to his or her own disaster recovery problems: Fault-tolerant software may provide a much-needed advance warning of disaster; modular software design can enhance the portability of applications from the disaster-stricken processor to a recovery host; and restoration of communications links between a recovered processor and user network may be expedited through the operational use of a particular type of channel extension equipment.

In short, if disaster recovery planning is integrated with other business planning and development projects, contingency plans need not be reinvented once a system conversion is an accomplished fact. The transition can be a gradual one that sees a revised disaster recovery plan available and in place just as new business systems come on-line.

When disaster recovery planning is first undertaken, planners almost invariably must "play the hand they are dealt." They must accept the vulnerabilities of existing business functions, systems, and networks, and do their best to reduce exposures. When a disaster recovery planner is directly involved in the initial development of a new business system or network, he or she is in an enviable position of being able to influence the shape of the subject system or network. This can only happen if disaster recovery planners are given the authority and the support of senior management to participate in other business planning endeavors.

Depending on the business organization involved, the authority may require a verbal order from a senior manager or an addendum to corporate policy. Whatever is required to position the disaster recovery planner within the project development team, this can be an effective means for reducing the lag between the time that a new business plan is implemented and a suitable strategy has been developed to safeguard it from catastrophic failure in the event of disaster.

PROJECT MANAGEMENT

Disaster recovery planning is a project similar to a systems development project. For this reason, it can be controlled and managed effectively using project management tools. Project management is defined generically as planning, organizing, directing, and controlling resources for a specified amount of time to meet a specific set of objectives. The formal techniques for managing a project have been the subject of numerous texts and are widely understood and employed by business people. These techniques include work definition and work timing; budget planning; resource allocation and scheduling; work progress tracking; and analyzing, forecasting, and correcting work performance.

A typical project will follow a baseline project plan, carefully described in tasks, milestones, costs, and resources. Complex projects may have many concurrent task paths; a large part of effective project management consists of identifying relationships between these concurrent tasks in order to define a critical path. *Critical path* is project management jargon for a simple idea: Some tasks provide necessary inputs to other tasks and must be completed before the other tasks may be initiated. Thus, effective project management may be viewed as the successful management of personnel and material resources so that critical tasks are realized on schedule and within budget.

In the real world, this sounds much simpler than it is. Project management techniques are the artifacts of effective project management. The stuff of effective project management is the same communication and motivational skills that are required of any manager. Perhaps because disaster recovery planning is often staffed by a loose coalition of personnel who are rarely dedicated full-time to the project, the communications skills supersede the skilled use of project management technique in realizing success in the disaster recovery planning enterprise. However, this book cannot teach communications skills. It can provide a lucid model for the project that a good communicator, but inexperienced disaster recovery planner, can emulate to achieve success in his or her planning project.

A final caveat must be added here. Some practitioners resist the categorization of disaster recovery planning as a project. The ongoing maintenance and support requirements of the disaster recovery capability, in their view, contradict the definition of a project as activity with a fixed time limit. However, systems development is typically undertaken under the rubric of project management and would need to be disqualified on the same didactic grounds were this criticism to be accepted. In fact, the disaster recovery planning project is complete once the plan has been tested, participants have been trained, and procedures have been established for identifying changes to business functions and their supporting systems and networks and maintaining the plan accordingly. When this has been done, the actual maintenance of a plan may become a job description rather than a project task.

SUMMARY

The material presented in this chapter has been largely introductory. The reader has been introduced to the concept of disaster recovery planning and has learned about the many activities that comprise a typical project.

Specifically, we have seen that:

1. Disaster recovery planning is an enterprise that is quite similar to other types of planning conducted within the company. As such, it is not a mysterious discipline accessible only to a group of specialists and experts. Those who possess the ability to communicate effectively and who are

versed in the techniques of project management have the prerequisite tools to develop an effective plan.

2. Disasters are interruptions of mission-critical business functions for an unacceptable period of time. The proper focus of the disaster recovery plan is not, as it has long been thought, the computer system, but all requirements for the restoration of critical functions including voice and data networks and a user community.

3. Disaster recovery planning is a pragmatic endeavor. So-called scientific risk analysis is a virtual impossibility given the multiplicity of variables and nonquantifiable exposures embodied in a disaster event. It is best to plan for a worst-case scenario that envisions total loss of company facilities. If a disaster having a lesser impact occurs, a well-designed disaster recovery plan may be implemented in sections to recover only affected systems, networks, and user operations.

4. Disaster recovery planning is a multiphasic process with internal feedback loops and complex activity relationships.

5. Senior management sponsorship is key to disaster recovery planning success. Typical rationalizations for management support are common sense, exposure management, audit considerations, and legal mandates.

6. Disaster recovery planning cannot assure recovery from a disaster. It establishes recovery strategies, provides logistical supports, and trains company personnel to deal rationally with an inherently irrational situation. There is substantial evidence that this, in itself, is valuable. Companies that prepare for the possibility of a disaster tend to weather them better than companies that do not.

7. An important activity within disaster recovery planning is the identification and acquisition of disaster prevention capabilities. This activity aims at minimizing or eliminating avoidable disasters and may also have the benefit of reducing company business insurance costs.

8. Disaster recovery plans are different for every company because they embody the specific requirements of each. To aid readers in orienting themselves to the disaster recovery requirements of their companies, the following checklist is provided.

Checklist of Disaster Recovery Requirements at My Company

1. Obtain a company organizational chart. Identify all work units (departments, product groups, etc.) and their managers. Plan to visit casually with each manager to learn more about his or her operations.

2. Identify what kinds of computer hardware are installed at the company. If you are unfamiliar with the operation of a particular system, make a note to speak with a knowledgeable user or system manager about its purpose.

3. Identify the telephone service provider for your company and find out the make and model of internal voice network equipment.

4. Begin researching disaster recovery cases in magazines and newspapers. Start a file, especially of case histories involving companies in your field.

5. Identify your company's business insurance provider. Find out the name of your local agent.

6. Discover whether disaster recovery planning has ever been attempted previously at your firm. A plan, if there is one, may be in the care of the company data center manager, records manager, or financial officer. Obtain a copy for review.

7. Identify local utility vendors and develop a list of contact names. Repeat this step for local law enforcement agencies, civil emergency management organizations, fire protection services, and medical facilities.

8. If your company is located in leased offices, find out the name of the leasing company and a leasing agent contact.

9. Discover what project management software is used at your company (if any) and obtain a copy. Learn its operations.

10. Read all you can find in print on disaster recovery planning. Send away for product literature.

11. Discover the name, location, and meeting times of any local contingency planning groups in your geographical area. A current list is provided in the next chapter, but new groups are forming at a quickening pace. Consult with computer user groups and data processing professional organizations to determine when contingency planning will be featured on their meeting agenda. Attend a session and talk to others who are engaged in planning for their firms.

3

Preliminary Steps

OBTAINING MANAGEMENT SPONSORSHIP

Senior management sponsorship is a key to disaster recovery planning success. Management's approval of the project is necessary to fund both the development of the plan and the acquisition of commercial products and services required by plan strategies.

Moreover, even in the smallest companies, there are often jealously guarded "territories" overseen by business or systems managers who may be less than receptive to the disaster recovery planner's data collection efforts. Senior management directives may be needed to surmount these obstacles and to obtain the cooperation of all parties.

In the best case, an enlightened senior manager will initiate the development of a disaster recovery capability. The manager's incentive may stem from a recent audit, or a new law, or even a lunch with an associate whose company has undertaken such planning. Often, the manager will turn to the company information systems executive with the task of developing a plan. The task may then be assigned to a subordinate within the company management information systems (MIS) or data processing (DP) area.

More often, the need for disaster recovery planning is first perceived within MIS or DP. Systems managers are exposed to disaster case studies and "war stories" in their professional journals and trade magazines and at meetings of their computer user organizations. They also tend to have a deeper understanding of the intricacies of restoring a complex system or network than do senior managers. Having performed many user needs analyses and developed many functional specifications to guide system development, they may have a greater appreciation of the impact on the company of even a short-term outage. Finally, they may take a personal interest in safeguarding *their* systems from harm.

Regardless of how the incentive to plan first materializes, it is often the case that a preliminary analysis—sometimes called a white paper or a feasibility analysis—must be undertaken to cultivate or cement management approval for the project. This analysis is less formal than a risk analysis and aims at developing a persuasive argument for initiating a formal planning project, rather than a full-blown definition of company exposures.

In this chapter, we will explore the common tasks involved in developing such a preliminary analysis. Once the analysis has been completed and presented to senior management for approval, we will look at several steps that may be taken to initiate the formal disaster recovery project.

Project initiation, as summarized in Figure 3.1, is the first major activity in disaster recovery planning. In addition to obtaining management endorsement, other initiation tasks include:

- Developing a project team and selecting a coordinator
- Developing an action plan, describing and assigning the tasks to be performed
- Developing a budget for the forthcoming risk analysis phase
- Establishing standards for documentation, software, and other tools to be used throughout the development of the plan

The way in which these preliminary steps are accomplished will have an important impact on all other aspects of planning. If only tentative support can be gotten from senior management, the successful management of the first planning phase—risk analysis—will be important to solidify management's confidence in the competence of the project coordinator and the viability of the enterprise as a whole. A professional and exacting approach must be taken to scheduling, budgeting, and controlling risk analysis, not only because of its intrinsic importance to developing an effective recovery capability, but also because of its value in consolidating the commitment of company decision-makers.

Also, the decisions made about such issues as who will lead the project, what software will be used to maintain electronic research files, and how the plan will be delivered, will ultimately determine the quality—defined as accuracy, scope, and maintainability—of the plan. Thus, these initiation activities are significant—though not very glamorous—components of a successful planning effort.

THE DISASTER RECOVERY PROJECT
PROJECT INITIATION

MAJOR ACTIVITIES

Conduct Informal Risk Assessment

Prepare Management Presentation

Obtain Management Endorsement

Develop Project Controls

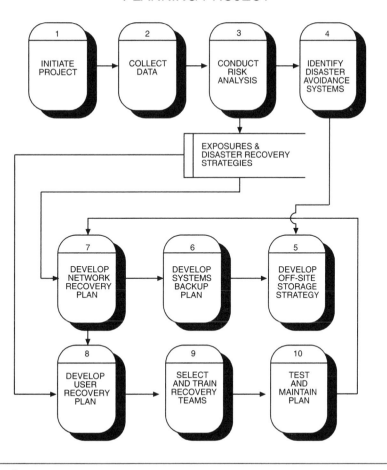

THE DISASTER RECOVERY
PLANNING PROJECT

Figure 3.1 The project initiation phase.

Conduct Informal Risk Assessment

The clearest case for disaster recovery planning resides in an assessment of the risk of disaster and of the corresponding loss potentials for the company. At the same breath it must be noted that it is difficult or impossible to establish a reliable statistic on the probability that a disaster will strike a company. Popular wisdom places the likelihood of a disaster at less than one percent. Some factors, however, may call into question this reassuring figure.

For companies located in hurricane- or earthquake-prone regions of the country, for example, statistics would certainly favor the possibility of disaster to a greater extent than for companies located in geographical areas that are relatively insulated from hazardous natural phenomena. The Federal Emergency Management Agency (FEMA) offers maps depicting certain types of natural and technological disaster potentials, and both the National Hurricane Center in Florida, and the National Severe Storms Forecast Center in Kansas City, Missouri[1] are sources of historical data on hurricanes, tornadoes, and other weather-related disaster potentials. Obtaining information from these sources can aid in strengthening the case for disaster recovery planning.

In addition to risk potentials from natural disasters, annual statistics are compiled and offered by the National Fire Protection Association (NFPA) covering equipment room fires that may be referenced to assess this disaster potential.[2] Similarly, historical disaster incident data may be available from industry or trade associations, from the Small Business Administration (SBA), and from disaster recovery product and service vendors that may be used in risk assessment efforts.

Not to be overlooked is the loss potential from computer security violations. Numerous law enforcement organizations and computer security professional groups track incidents involving computer crime, many of which result in system or network interruption. There are at least a dozen computer security periodicals and a host of federally supported studies that may be reviewed to obtain data on this risk category.[3]

Other local sources of information on risk potentials may include local civil emergency planners, utility companies, telephone serving offices, and even the company's building manager (assuming facilities are leased). Understanding how power and telephone service are provided to your facility, who your neighbors are and whether their business activities pose potential hazards, and how civil emergency planners regard the company's location (is it in flood risk area or a hurricane evacuation zone, for example) may identify other risk potentials and support an assessment about the possibility of a prolonged business interruption.

---

[1]To obtain information on potential natural disaster risks in your geographical region, write to: FEMA, Office of Risk Assessment or Office of Natural and Technological Hazards, 500 C Street SW, Washington, DC 20472; National Hurricane Center, 1320 South Dixie Highway, Room 631, Coral Gables, FL 33146; National Severe Storms Forecasting Center, 601 East 12th Street, Room 1728, Federal Building, Kansas City, MO 64106.

[2]Write to the NFPA at Batterymarch Park, Quincy, MA 02269.

[3]Contact the National Institute for Standards and Technology, US Department of Commerce, Gaithersburg, MD 20899, and the Information Industry Association, 555 New Jersey Avenue NW, Suite 800, Washington, DC 20001, for information on private and public computer security studies.

The culmination of this research need not be synthesized into a probability statistic. It may suffice, in an informal risk assessment, to identify the risk potentials in a few sentences each, noting impressive frequency statistics or significant quotations from experts or dollar losses attributed to recent cases. In this way, management may be advised of the possibility of disaster and the need for planning without treating the issue of what the chances are that a disaster could occur.

Gather Industry Statistics

Once disaster potentials have been identified, it is necessary to develop an estimate of how much the company stands to lose if a disaster strikes. Here again, the effort should not be (as it is in the formal risk analysis) to provide a meaningful dollar-cost figure. It will suffice to identify, in general terms, that the company stands to lose a great deal in the event of disaster.

There are two components to this effort. First, gather industry statistics to identify how other companies in the same line of business would fare in a disaster. The second component is to research case histories of disasters involving companies whose business interests are similar to those of the planner's firm.

Gathering industry statistics is a time-honored method for placing company needs in perspective. Fortunately, most disaster recovery planning studies categorize their survey respondents by industry type. Through these academic studies, the planner may be able to learn such trivia as the maximum amount of downtime that may be sustained by a business within a particular industry before recovery is impossible. In the case of one industry standard survey, statistics are also provided to show the average dollar loss accruing to the company (again, subdivided by industry group) for 24, 48, and 72 hours of downtime.

Additionally, trade associations, business insurance companies, and some disaster recovery service vendors have conducted surveys to develop an understanding of industry-specific disaster recovery vulnerabilities and preparedness. With some digging, the disaster recovery planner may be able to locate this information and use it as a comparison estimate of his or her company's exposure.

Again, only relevant industry-specific data need be presented to management. The planner need not be exact about the company's exposure and should be quick to point out that such an estimate is impossible to make without formal risk analysis. There may be value, on the other hand, in listing some of the costs that will accrue to the company without assigning dollar values. These will include:

- Equipment replacement costs
- Personnel overtime
- Delayed accounts receivable
- Delayed payroll
- Missed sales opportunities
- Delayed order fulfillment
- Loss of customer confidence
- Facility damage and salvage costs
- Application software losses

The list may become quite extensive based on the business functions of the company. A just-in-time manufacturer, for example, may quickly lose clients if orders cannot be processed and shipped immediately upon receipt. Banks face the possibility of crisis withdrawals or balance of payments crises that may develop very quickly. The more specific that the planner can be about the loss potentials that the company faces, the more realistic the preliminary analysis will seem to senior management.

In addition to surveying industry statistics on dollar loss potentials, the planner should also note data on industry disaster preparedness. Some studies indicate the percentage of respondents reporting that they have undertaken some kind of disaster recovery planning. If these numbers are meaningful in the planner's industry group—if, for example, they show a trend toward planning—they could be cited to bolster the case for disaster recovery planning at the planner's company.

Research Case Histories

At one time, information about disasters was very difficult to obtain. Businesses, understandably, were reluctant to publicize their misfortunes, due in part because they feared a loss of client or customer confidence. It was said that the best information about disaster recovery was locked away in the personnel filing cabinet in the exit interviews of those who were blamed for the event.

Today, there is regular reportage on disasters by the news media and in the computer trade press. Companies are becoming more savvy about media and are learning that forthrightness is generally a good policy for obtaining media support. Also, a disaster from which the company recovers due to effective disaster recovery planning can actually become a media asset, gaining notoriety for the recovered firm as "unsinkable" or "effectively managed" by business people who protect their operations and provide for their customers.

The result of this development is an increasing library of case studies on disaster recovery. Nearly every industry now has one or more *archetypes*—that is, disaster recovery incidents involving companies within the same industry—from which they can learn much about industry-specific disaster recovery requirements.

The planner should scour industry and computing journals to identify relevant cases and collect as much information as possible about them. Also, if an article quotes individuals who participated in the recovery—both from the affected company and its disaster recovery service vendors—planners should follow these leads with telephone calls to obtain information that may not have appeared in the press. For example, typically the media will cover a disaster soon after it has occurred. Invariably, a loss estimate will be provided by an interviewee. Actual loss data may not be calculated until long after the event and by then it is no longer "newsworthy." The planner may be able to obtain this information, as well as some lessons of the disaster identified by recovery personnel in hindsight, by making personal contact with recovery personnel.

A case study can be a valuable tool for communicating to management not only the scary reality of disaster, but also the demonstrated benefits of effective planning. The emphasis, in fact, should always be on the latter.

Identify Relevant Legal & Regulatory Mandates

As previously noted, certain legal and regulatory mandates, both federal and local, may compel management to undertake disaster recovery planning. Chances are very good that auditors have already advised management of any legal requirements.

However, some requirements are couched in terms of computer security and data integrity. The disaster recovery aspects of these requirements may not be as familiar as the requirements for data encryption, backup and storage, and physical access controls.

It is valuable, therefore, to mention the law, but planners should be careful not to appear to lecture or otherwise overstate the legal mandate. Rarely will the absence of a disaster recovery plan become an issue unless the company is being prosecuted, audited by the IRS, or sued on some other grounds. Also, except in the case of federally insured banking and savings-and-loan institutions, the requirement for disaster recovery planning is not very clear-cut—and even many of these institutions appear to be satisfying legal requirements with the bare minimum of planning.[4]

Develop Justification

The justification for undertaking disaster recovery planning is derived from the information previously summarized. At this point, the planner can only point to the following facts:

1. The company is exposed.

2. Industry exposures (as defined by survey research) average X number of dollars and the costs of a disaster will increase as the duration of interruption increases.

3. Company XYZ (based on a case study) sustained X dollars of loss from an outage lasting X hours, but recovered due to effective disaster recovery planning.

4. In addition to the business factors supporting disaster recovery planning, there may also be legal or regulatory requirements that support undertaking the plan.

The planner should also be clear that what is being sought is not a blank check to develop a disaster recovery plan for the company. What he or she hopes to justify is the expenditure of some monies to conduct formal risk analysis in order to identify the current level of preparedness at the company, to identify company-specific risks and exposures, and to develop recommendations for developing a plan that will provide for the safety of employees and the continuation of business operations in the wake of a disaster.

---

[4]Even in the case of governmental computing, where regulatory requirements are quite exact, there seems to be little enforcement and only the smallest of budgets for developing and for auditing disaster recovery plans.

Conduct Management Awareness Briefing

This justification may be addressed in a memo, but it may be effective to supplement the document with a briefing. This provides the means for senior management to air concerns or to ask questions and receive answers—interaction that is not possible with a paper argument.

The briefing should be held at management's convenience. (It probably wouldn't occur at any other time!) Effective briefings will consist of:

- **Advance planning**—Evidenced by the preparation of an agenda, the use of visual aids, the reproduction of an adequate number of handouts for meeting participants (the planner should make a few more copies of the memo than are needed), and the reservation of needed meeting rooms, presentation equipment, coffee, and so forth.

- **Brief, concise presentation**—Planners should rehearse their presentations and keep them under 10 minutes, which is the average adult attention span. Above all, it should be made clear that what is being sought is authorization to proceed with a risk analysis, in order to define company needs more exactly. Keep technical terms and language to a minimum and focus on business concepts such as operations and exposures, not on academic concepts such as methodology, functional data collection and analysis, or platforms and configurations.

- **Facilitative question-and-answer**—Planners should field whatever questions are asked to the best of their ability. The need for disaster recovery planning should be stated but not belabored. Planners should concede a lack of information in questions that they cannot answer, but include an offer to research the answer and return a response at a later time.

- **Positive attitude**—The goal of the management briefing is to advise, not to sell. Keeping this in mind, the planner should not communicate missionary zeal, but professional interest. The justification that is being offered is thin at best. Planners must rely on management to consider the case that has been presented and to decide whether it merits further action.

Some planners are pleasantly surprised to learn that management has been considering the need for disaster recovery planning for some time, but has not moved forward due to other more pressing operations issues. They may be receptive to the planner's ideas, assign a manager to be the point of contact on further discussions, and express their desire to proceed with the development of a plan.

On the other hand, the planner may be dismissed with perfunctory thanks and advised much later (if at all) of their reaction to the presentation, positive or negative. If authorization is not forthcoming at the presentation, the planner should draft a memo to meeting participants, thanking them for the opportunity to present the proposal and expressing desire to perform the risk analysis task on behalf of the company. If there is no response after some time has passed, the planner may need to follow up with a memo inquiring as to the status of the proposal, possibly including some timely information about a new case study or survey.

If all else fails, the planner should seek the sponsorship of a manager who is receptive to the idea of disaster recovery planning and who has ongoing access to senior management. If a manager will champion the proposal, and keep it in front of management, a decision will ultimately be rendered.

Recommend Corporate Policies on Contingency Planning

Assuming that management agrees to the underwrite the risk analysis, this support should be consolidated as quickly as possible. There may be obstacles to conducting data collection if departmental and systems managers are not aware that the project has senior management endorsement; they may prefer not to be bothered with the intrusion of an "outsider" on their "turf."

Perhaps the simplest method to avoid this difficulty is to draft a company policy on disaster recovery planning and to submit it to management for approval. Such a policy statement may even be included in the initial presentation on the need for disaster recovery planning. The policy recommendation need not be lengthy. It should state that all managers share an interest in providing for the safety of personnel and the continuation of business operations in the event of a disaster and should comply with efforts that are being undertaken by senior management to develop plans for reacting to and recovering from a disaster.

In some cases, policies are much more exact and legalistic, stating that the company will develop a plan to provide for the recovery of key business operations within X hours following a disaster; that the plan shall be tested at intervals at least X times per year; that tests will be attended and witnessed by X personnel; that employee awareness and safety programs will be initiated and training will be conducted X times per year; and so forth. At this early stage, however, such a policy would be difficult to define and unlikely to receive management's nod, unless they are predisposed to undertake planning or initiated the project in the first place.

Typically, any policy statement endorsing disaster recovery planning delights auditors and regulatory oversight personnel. Management may be willing to adopt a policy statement on these grounds alone. If management is reluctant to give consent, the planner should ask at least for a memo to be directed to department heads advising them of the task that is about to be undertaken and identifying the planner as the person responsible for the task.

Support for risk analysis and plan proposal development must be obtained from senior management for disaster recovery planning to proceed. However, planners may also profit from the cultivation of support among middle managers. Simply put, the more support and interest expressed to senior management by department heads, the more likely the disaster recovery plan project is to be funded and completed. To cultivate support, the planner should:

- Explain the rationale for planning at every opportunity.
- Commend any efforts that may have been taken by managers to back up data or records.
- Reassure managers that disaster recovery planning is in their interest and that their attention to the need for such planning reflects positively on their ability as enlightened managers.

Planners should keep in mind that the needed data on departmental operations is exacting. Collection will be time-consuming and will require the assistance of department personnel. Whatever good will can be developed should be encouraged, as it will be taxed when data collection efforts begin.

ASSEMBLE PROJECT TEAM

Disaster recovery planning is a team effort. Even if the planner is the only individual with full-time responsibility for the effort, others—including technical, clerical, and managerial personnel—will be involved. (Even in cases where a consultant is called in to perform disaster recovery planning, there is typically a plan coordinator assigned by the company to work with, oversee, and/or learn from the consultant.)

Considerations for Staffing the Project Team

Ideally, the project team will consist of the planner, a representative of each corporate function or department (including internal audit), and technical experts as required by the company's use of computer and network technology. Clerical personnel, perhaps a project secretary, will also be assigned to the effort to record meeting minutes, coordinate meetings, and so forth.

The planning team must remain small to be effective and to keep project labor costs to a minimum. Also, team members should be reliable, punctual, and—if they are serving in other capacities within the company—should have sufficient time in their schedules to perform the tasks that are demanded of them.

Management

This book envisions the project manager—also referred to as the planner or coordinator—to be a company employee. This may seem to contradict the experience of many companies that develop their disaster recovery plans using outside consultants, but this contradiction is illusory.

Consultants may be viewed as potential members of the planning team, and as technical experts on the basis of their experience with disaster recovery plan development. However, even when an outside consultant is employed by a company, the consultant generally operates under the direction of and/or in conjunction with a company planner.

The question of who should develop the plan—in-house personnel or outside consultant?—is becoming increasingly irrelevant as better information about the techniques of disaster recovery planning is becoming available to anyone who wishes to gain this skill. The new question is whether a consultant is necessary at all or whether the hiring of a consultant is a superfluous expense?

Figure 3.2 provides a decision-making matrix for use in evaluating the case for obtaining disaster recovery consulting services.[5] From this matrix, it can be argued that consultants are valuable from the standpoint of their experi-

---

[5]This matrix was originally published in Jon William Toigo, "The Disaster Recovery Plan: Who Should Develop It?" Portfolio 85-01-30. *Data Security Management* (New York: Auerbach Publishers, 1988): 11.

ence with several companies, when plans must be developed quickly, or when technical complexity is an issue. Moreover, the expense of a consultant can sometimes be leveraged to surmount the reluctant participation of middle managers, or to cultivate management confidence in planning performance.

An in-house development effort is favored over the consultant option based on consultant expense, when technical complexity is offset by in-house technical capabilities, and when senior management support is forthcoming. Of course, the in-house disaster recovery coordinator must also possess the requisite skills to develop the plan or have access to the resources required to develop those skills and be permitted to work on the plan as a full-time job.

Having said this, the position of this standard should be reiterated: Any employee having prerequisite communications skills, a basic grasp of disaster recovery planning objectives, and a background in project management, can develop a disaster recovery plan that may be tested and refined so as to provide a recovery capability for his or her company. From this perspective, consultants are viewed as a potentially useful technical resource for disaster recovery planning, but not as a substitute for effective internal plan management.

Now, it is time to examine the disaster recovery project manager's role in greater detail. Within the context of project initiation, the planner's managerial capabilities are first put to the test.

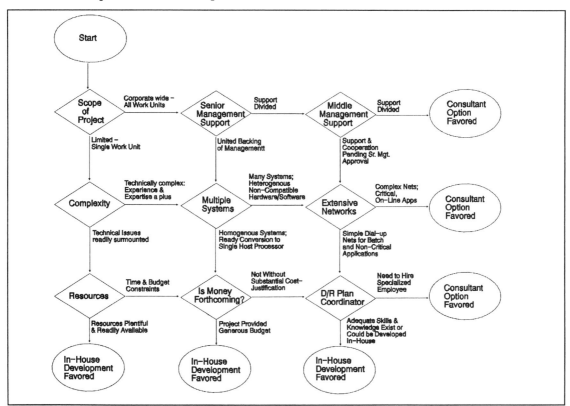

Figure 3.2 Evaluating the consultant option.

Facilitative Leadership

Given the ad hoc nature of the disaster recovery project team (and, by definition, of all project teams), effective management translates to facilitative leadership. The planner, with his or her background in project management and knowledge of disaster recovery planning, provides the vision for the disaster recovery project. This vision must be communicated to other team members in terms other than commands and directives. Thus, a facilitative model for interaction with the team is the best model.

What is facilitative leadership? It has many dimensions, but at its core is a respect for the experience, knowledge, and skills of individuals as represented in their opinions. The disaster recovery planning team may be comprised of individuals with diverse backgrounds and psychologies. A facilitator manages the interaction of these individuals to enhance cooperation and minimize conflict.

Technical team members are to be appreciated for their refined knowledge of the technologies supporting business functions. They will have the task for developing technical solutions to system and network recovery. However, these technical solutions do not supersede the needs and interests of business managers who are represented on the team. Business managers understand the work of the departments within the company and probably have a better appreciation of the interrelationships between business functions, the costs of operations, and of the marketplace. This background is equally vital to establishing a plan for the recovery of the business.

Without an appreciation of these differences, a planner—depending on his or her own background—may favor one group over another. If there is one truism, it is that business-minded team members will be of enormous assistance in identifying recovery priorities and technically-minded team members will support the business priorities with technical solutions. To capitalize on these strengths, the planner must give both groups their due, serve as an honest broker to negotiate disagreements, and ask questions more than deliver edicts—in other words, the planner must be a facilitator.

Budgetary and Administrative Functions

The disaster recovery planner, as the manager of the disaster recovery project, is responsible for budget and administration. This responsibility entails:

- Formulating a budget; approving expenditures; coordinating with company payroll, accounting, and purchasing functions to obtain needed goods; and recording and documenting expenses.
- Interfacing with senior management to present, justify, and review project plans; to report progress and problems; and to solicit support for planning activities and expenditures.
- Interfacing with department managers and system administrators to obtain their cooperation and assistance in data collection, backup policy implementation and enforcement, plan testing, and installation of disaster prevention capabilities, and working with managers and administrators to resolve conflicts.

- Selecting and recruiting project team members and working with company human resources and department managers to coordinate work schedules, time billing codes, and other personnel issues.
- Monitoring team performance; assigning tasks and responsibilities; evaluating work quality; training; and, discipline.
- Negotiating disputes, building team solidarity, and serving as advocate and ombudsman for the team.

Project Management Skills

The planner as project manager has additional responsibilities that relate to the project itself. Specifically, the planner must:

- Establish the scope of the plan, articulate objectives, and define subordinate planning tasks and milestones.
- Establish standards for work performance and project tools.
- Define work components, required resources, and identify a critical task path.
- Schedule work, monitor schedule adherence, identify and surmount obstacles to scheduled work.
- Maintain accurate records of work including a baseline project model and actual data; identify variances and their causes; and make periodic reports to senior management or their designated representative on progress.
- Validate or verify completed work.
- Establish acceptance criteria for project deliverables and manage the approval process.

The planner should possess a background in the techniques of project management as embodied in the previously cited tasks. Plus, he or she should be skilled in the use of automated project management software packages that are employed at the company (if such packages are used). The use of an automated project management tool may provide the planner with additional time to participate directly in data collection efforts and strategy formulation.

Technical Experts

Effective disaster recovery planning requires the coordination of the talents and skills of numerous experts. As previously noted, consultants may be hired who provide experience in developing plans for other companies and extensive knowledge of the disaster recovery industry. Company personnel who are well-versed in system and network technology and who have participated in the design of company applications may be utilized to aid in the development of strategies for recovering those capabilities in the event of a prolonged interruption. Business managers or system users may be recruited to add a much needed perspective on business function criticality and business operations and standards. Company auditors may be invited to participate and contribute their expertise in plan justification, quality assurance, and testing.

In the following sections, we will examine the roles and responsibilities of these potential players in the formulation of the company disaster recovery plan.

Consultants

As previously stated, the disaster recovery consultant can offer useful insights into the planning project based on experience in developing plans for other companies and knowledge of the products and services of the disaster recovery industry. Consultants, however, are often expensive. Therefore, their use should be carefully considered and maximized.

Figure 3.3 provides a checklist for evaluating consulting services. To make the most of consulting services, one must identify the right consultant for the job. Preferably, the consultant will offer experience in developing disaster recovery plans for clients that are in the same industry as the host company. Client references should be requested and verified.

---

1. Verify that the consultant whose services are being acquired is experienced in developing plans for other companies in your area of business.

2. Call consultant reference accounts and discuss past performance. Learn what services the consultant provided. Ask for personal opinions on performance:

   Was the service effective?
   Did the consultant set and meet schedules?
   Did the consultant recommend other disaster recovery products and services? Which ones?
   Did the service justify the cost?
   Does the consultant maintain the plan?
   Does the consultant participate in plan testing?
   Did the consultant provide training for plan participants? For an in-house plan coordinator?
   What does the past client regard as the consultant's greatest contribution?
   Would the past client hire the consultant again?

3. Identify areas in which the consultant would be most useful:

   Evaluating data collection questionnaires?
   Aiding in Risk Analysis?
   Reviewing results of Risk Analysis?
   Formulating recovery strategies?
   Developing the disaster recovery plan document?
   Evaluating the paper plan?
   Developing training programs for plan participants?
   Developing testing strategies?
   Analyzing test results?
   Maintaining the plan?

4. What is the consultant's methodology? Is it consistent with a project management approach? Does the consultant use an automated disaster recovery planning tool?

5. Ask for a road map of the disaster recovery planning project, depicting the consultant's approach.

6. What are the consultant's rates?

7. Does the consultant have any relationships with vendors of disaster recovery products and services?

Figure 3.3 Consultant checklist.

In addition to consultant pricing, the planner should investigate the consultant's relationships with vendors of disaster recovery products and services. If the consultant will be called upon to participate in recovery strategy formulation, it would be useful to know with which vendors the consultant has co-marketing arrangements—both as a check for conflict-of-interest and as a means of identifying pricing advantages that may accrue to the use of a particular consultant.

Many disaster recovery consultants offer full-service support for disaster recovery planning, including an automated disaster recovery planning tool, a methodology, and even a boilerplate plan. These should be reviewed for their consistency with the company's project management approach. If the planner prefers to use the consultant only in certain areas of plan development, these should be identified and time requirements estimated. Then, a proposal should be solicited from the consultant for work charges.

Consultants can be an asset to disaster recovery planning if used properly. Experience, the major contribution of the consultant, is best utilized in risk analysis, plan document design, and testing strategy formulation. However, even these activities may be adequately performed by in-house personnel, given budgetary constraints.

Systems and Network Specialists

Disaster recovery planning, in large part, consists of building emergency systems and networks to provide critical services which have been disrupted on their normal platforms. Those who design, implement, and maintain systems and networks within the company are often the best sources of information on how to restore those systems and networks in the event of disaster. For this reason, most technical expertise required for disaster recovery planning should be sought from in-house technical personnel.

As documented in later chapters, the tasks confronting these team members include the following:

- Development of a minimum acceptable equipment configuration to provide critical data processing services for a designated period of time.
- Development of a minimum acceptable network configuration to handle critical voice and data communications requirements for a designated period of time.
- Formulation of software design standards to facilitate application software portability from host systems to the minimum acceptable equipment configuration.
- Collection and analysis of usage statistics, documentation of on-line and batch applications, documentation of application security features and solutions, and identification of logistics requirements for remote site operations.
- Development and implementation of a large systems data backup strategy.
- Development of guidelines and programs for micro-computer and departmental minicomputer backup and safety.
- Recovery strategy testing and maintenance.

Company technical experts will be required to provide solutions for data processing and networking requirements in a disaster situation. They will use standard design and development techniques—commonly employed for new systems development—to create a system that will meet the needs and objectives of the company as defined in the disaster recovery plan.

Business Function Experts

If company technical experts will work to develop recovery systems and networks, business function experts—managers or staff representatives of key departments of the company who have been recruited to the disaster recovery team—will identify the requirements that the emergency systems and networks must meet. A business function expert will examine each department of the company, identifying typical work processes and flows. They will solicit from the personnel of a department their opinions about the impact of a disaster and the ability of the department to do without automated systems or networks.

Having collected this information, as well as data on terminal usage, records retention, and other work requirements (FAX, photocopy equipment, microfiche readers, preprinted forms, etc.), the business function expert will endeavor to apply a criticality classification scheme to the functions of each department. Functions that are identified as critical, together with their personnel, system, network, facility and logistical requirements, will become recovery objectives for the disaster recovery plan.

Auditors

The role of the internal auditor in disaster recovery planning is much the same as it is in normal business operations. The auditor is concerned with legal and regulatory compliance, with data integrity and security, and with quality assurance. Not only can the auditor help to make the case for disaster recovery planning, he or she can provide other useful roles as well, such as:

- Identifying business functions which are critical, not because of their contribution to continued business operation, but for legal or regulatory compliance.
- Aiding in the formulation of plan testing strategies using tools commonly employed in conventional EDP audits.
- Overseeing the test evaluation process and ensuring plan maintenance.

Additionally, auditors, who serve as the "eyes and ears of senior management" can provide an excellent channel for keeping management informed of plan progress and obstacles, for recommending strategy funding, and for justifying plan proposals.

Auditors may also be of assistance in evaluating insurance coverages and identifying areas in which premiums could be reduced based on disaster recovery provisions.

Clerical and Support Personnel

Disaster recovery planning is a complex endeavor with large information collection, processing, and distribution components. Meetings need to be documented, data collection interviewing schedules need to be coordinated with project heads, reports on project status must be generated, numerous solicitations for vendor bids must be issued, responses received, and proposals documented.

Some disaster recovery project managers utilize the services of a project secretary to fulfill these tasks. In some companies, the word processing pool may be called upon to aid in some of the deliverable preparation work. Other resources may be required on an intermittent basis.

For the sake of budget, the project manager should strive to keep these costs to a minimum, possibly requiring team members to perform their own typing chores and assigning meeting secretary chores to team members on a rotating basis. However, it is also important not to short-staff the project and risk becoming swamped in paper. At least one full-time individual should be requested to handle the enormous flow of information generated in a disaster recovery project.

DEVELOP AN ACTION PLAN

The first meeting of the disaster recovery team is used to introduce team participants, define the scope of the project, identify general objectives, assign tasks and responsibilities, schedule preliminary tasks, and establish standards for the project. This is an enormous amount of ground to cover when team members are meeting for the first time, so it is valuable to establish an action plan for distribution during or prior to the meeting. This section will examine the components of the action plan.

Define the Scope of the Plan

The scope of the plan corresponds roughly to its master objective. The disaster recovery plan seeks to provide a testable capability for continuing business functions in the event of a disaster.

The project team should understand that:

1. The plan will encompass business functions throughout the company.

2. The plan will be developed as a project and they will participate as a project team.

3. The plan will be developed in phases, with a management approval step at the end of each phase.

Identify General Objectives

To define general objectives, the planner will need to first review the overall model for the project. Objectives are best stated in terms of deliverables.

The list of common deliverables of disaster recovery planning is provided in Figure 3.4. Subsequent chapters will provide samples of these deliverables and instructions for their use. The general objectives underpinning these deliverables are:

1. Obtain senior management approval to initiate disaster recovery planning.

2. Perform data collection on business functions and document results.

3. Perform data collection on systems, networks, and document results.

4. Identify and quantify loss potentials accrued to an interruption of mission-critical information services and determine maximum allowable downtime.

5. Classify business functions according to criticality and establish priorities for business function restoration in the event of an interruption.

6. Formulate strategies for systems, networks, and user recovery that will restore critical business functions within maximum allowable downtime limit.

7. Institute a company disaster awareness program.

8. Institute a data backup and off-site storage program and secure such additional disaster prevention or containment capabilities as are approved by management.

9. Develop and implement a training program to familiarize those who will play a role in plan implementation with recovery tasks and procedures.

10. Develop a disaster recovery test plan, conduct tests, and evaluate results.

11. Institute procedures for collecting information required in periodic plan maintenance.

---

0. Develop a Preliminary Risk Assessment.
1. Develop a survey instrument for collecting data about business functions.
2. Develop a survey instrument for collecting data about end-user terminal usage.
3. Conduct a risk analysis and formulate a Risk Analysis Summary.
4. Develop a criticality spectrum.
5. Assign criticality values to business functions.
6. Develop a Business Function Restoration Priority Position Paper.
7. Formulate a Systems Recovery Plan.
8. Formulate a Network Recovery Plan.
9. Formulate a User Recovery Plan.
10. Develop a Bid Evaluation Worksheet to compare vendor offerings.
11. Design a Disaster Recovery Plan document.
12. Develop an Emergency Decision-making Flowchart.
13. Develop corporate disaster awareness program.
14. Develop and administer disaster recovery training curriculum.

Figure 3.4 Deliverables checklist.

15. Develop Disaster Recovery Test Parameters document.

16. Develop Disaster Recovery Test Evaluation form.

17. Develop Disaster Recovery Maintenance Record form.

18. Develop Disaster Recovery Maintenance Advisory form.

Figure 3.4 Deliverables checklist. (Continued)

To these general objectives may be added others determined by the specific needs of the planner's host company.

Allocate Tasks and Responsibilities

The immediate action items for the first meeting will consist of developing survey instruments for data collection in company departments and assigning team members data collection targets. These items must be accomplished to move to the next stage of plan development—risk analysis.

The sample data collection forms provided in Figures 4.1 and 4.2 may serve as a starting point for the development of customized collection instruments (see Chapter 4). The planner may wish to copy and distribute these models in advance of the meeting so that participants can review them and note recommended changes prior to the meeting.

In addition to the data collected through interviews, information on existing systems and networks—including flow diagrams, specifications and hardware/software lists, network diagrams, and functional descriptions of applications—needs to be assembled. Technical members of the planning team should work to compile a list of available and required system and network documentation at the first meeting. Additional documentation that is required but unavailable should be assigned to the appropriate team personnel.

Next, using an organization chart, responsibility for conducting interviews with designated end users should be assigned to team members. Instructions on the proper methods for conducting interviews and procedures for the proper application of survey instruments should be discussed before the meeting adjourns and until the plan coordinator is confident that each team member understands his or her assignment.

It may also be beneficial to use part of the initial meeting to review the list of general objectives and to assign responsibility for these tasks to individuals or groups. Doing so will facilitate advance work that team members may need to perform, such as soliciting vendor literature or researching relevant writings on techniques or technology.

Schedule Task Completion Dates

Target completion dates should be established by the group for data collection and for subsequent objectives. These dates will be carefully recorded to provide a baseline schedule for the plan.

Formalize Data Collection Tools

Figure 4.1 (see Chapter 4) provides a model for use in developing a business function survey instrument. The instrument is used to obtain information about business functions of the company, including the purpose of the function, typical procedures involved, requirements of the function, and system and network components. Also sought are the opinions of departmental managers or supervisors regarding the criticality of the function, how they might manage if their systems or networks were unavailable for a prolonged period of time, and what they would calculate the cost to the company to be of a prolonged interruption.

Figure 4.2 (see Chapter 4) provides a model of a terminal usage profile form for use in collecting information on computer and telecommunications equipment employed in work performed by the department. This form may be adapted to suit the needs of the reader's disaster recovery planning effort.

Formalize Analytical Methodology

Data collection instruments will be used to assemble information about the business for subsequent use in determining criticality and restoration priority. Data on downtime cost will also be used to cost-justify subsequent planning activity to senior management.

The analysis of data is covered in detail in Chapter 4. However, it is helpful for the project leader to provide the team with a description of risk analysis requirements that will aid them in getting the right data from interviewees. The planner should review Chapters 3 and 4, then summarize key considerations for presentation to the team.

Develop Formal Meeting Schedule

Data collection and subsequent planning activities are complex and time-consuming. Difficulties may develop in collecting needed information, dealing with reluctant departmental or systems personnel, and in many other areas. To keep the project on schedule, these problems and issues must be quickly brought to the attention of the team and the project coordinator so that obstacles can be surmounted. Regular weekly or bi-weekly meetings will help to keep the project on track. Obtain the opinion of the group as to the best time and day for regular team meetings.

COORDINATION AND CONTROL

Before the initial meeting of the planning team concludes, it is important to introduce controls that will help to ensure that the products of many individuals may be integrated into a whole. These include documentation standards, software standards, and procedural controls.

Documentation Standards

The use of prepared forms for data collection and other planning activities helps to reduce the time requirements for collating the information of numerous personnel. The importance of utilizing these forms should be stressed to team personnel and survey respondents.

Other deliverables and documents of the project, including bid specifications, personnel requests, expense reports, and the paper plan itself, may need to be developed in accordance with company documentation standards. The plan coordinator or the project secretary should be responsible for identifying relevant standards and for ensuring that deliverables conform to them.

Software Standards (Generic Applications)

Certain software tools—including word processing, database, project management, and analytical or spreadsheet applications—may be used in the development of project deliverables. Standards need to be developed to ensure that data produced by project participants can be utilized by other team members.

Word Processing

Word processors will be used extensively by team members in developing task deliverables. Individual preferences should be accommodated to the extent possible to avoid the time delays accruing to learning an unfamiliar product. However, all packages should be capable of saving files in a format compatible with all word processors that are in use.

A generic standard such as a corporate word processing file format or ASCII text should be articulated as the standard file format. Also, data portability should be provided by standardizing on a common diskette format or by some other means. If a centralized processor to which all team members have access is available within the company, this may serve as a repository for data files.

Periodic backups of project data should be made according to a defined schedule and removed to safe storage. Responsibility for backups, if this is not accomplished via normal business systems operations, should be assigned to a team member.

Project Management

As previously noted, the project leader may find useful an automated project management software package for documenting and managing plan development. A company-standard product should be used or the team leader should select the package with which he or she is most comfortable.

The output from the project management package should be backed up regularly and removed for safe storage. (This point, which will be reiterated in all software standards found in this book, is extremely important. It is bad form for the disaster recovery planning project to be delayed due to a disaster that claims project data for which there is no backup!)

Should other team members have need for a project management application, the planner should require that they use a package that is file-compatible with the master project management package. This is a hedge against the possibility of personnel turnover with its consequences for data access.

Database

Database application software provides an effective means for storing and managing list data (though this can also be done using most high-end word processors). Hardware inventories, personnel telephone contact directories, and even survey control data may be managed using a database.

In most cases, however, developing a customized database is a time-consuming activity. It is also potentially fraught with difficulty should the database author leave the company. These potential drawbacks can be minimized by ensuring that facilities exist for dumping the contents of the database to an ASCII text or other file format that may be imported into a word processor. If database application software is used by team members, ensure that data files are backed up at regular intervals and removed to safe storage.

Analytical and Spreadsheet

Spreadsheets are excellent tools for recording budget information and for documenting the costs of proposed recovery strategies. They may also be used to manage list data.

Most current microcomputer-based spreadsheets will export files to popular word processor or ASCII text format. Object Linking and Embedding (OLE) in the Microsoft Windows environment allows Windows-compliant word processors, spreadsheets, and graphics applications to share their data with no conversion requirements. Planners should ensure that requirements for file sharing or conversion can be satisfied before acquiring a spreadsheet application. They should also be sure to include spreadsheet files in a regular program of data backup and secure storage.

Software Standards (Special Applications)

In addition to generic software applications, disaster recovery planners may wish to examine application software that is marketed strictly for use in corporate contingency planning. These specialized packages are of two general types: risk analysis and disaster recovery plan development.

Risk Analysis

Risk analysis software provides an automated facility for recording information about business functions, operating costs, and producing estimates of dollar exposure. Novice planners, who have been told in vendor literature that risk analysis is an extremely difficult process, may find the prospect of relying on an automated package desirable. However, planners should be aware of the following aspects related to using automated packages.

1. Data input to a risk analysis package must be collected first. Collection of reliable data is the most difficult aspect of risk analysis and is not aided by these software packages.

2. Data manipulation by the risk analysis package reflects the opinions of the software author regarding the relative importance of collected data. This weighting system may not be consistent with an objective assessment

of business functions. To determine whether this is the case, one would need to perform a risk analysis by hand—which is what the packages are supposed to help planners avoid.

3.   Data output by an automated risk analysis package may contain hidden premises that planners would be hard-pressed to explain or justify if questioned by senior management.

Should planners decide to review risk management software for possible adoption by the project, they should keep these possible drawbacks in mind. They should also be assured, after reviewing Chapter 4, that risk analysis is an inexact science and that the best risk analysis is the product of a pragmatic review of quantifiable and nonquantifiable loss potentials.

With the emergence of expert systems, there is a possibility that a risk analysis product will be offered some day that will produce a worthwhile portrayal of company exposure. Thus, it is in the interests of the reader to review product packages and to keep abreast of changes in this software category.

Disaster Recovery Planning Software

The subject of disaster recovery planning tools has been addressed previously. Software in this category comes in two basic types: word processing-based and database-based. Essentially, vendors offer a centralized tool for recording company information and outputting a paper plan. Some have added risk management facilities, expert system drivers, and automatic dialing directories.

Without belaboring the point here, it should be noted that *canned plans* almost invariably stress data-center recovery over corporate-wide recovery, and that they frequently require the adoption of a particular planning methodology that may or may not agree with the effective project management philosophy expressed in this book.

Should the planner wish to evaluate these packages, a checklist is provided in Figure 3.5 for use in this endeavor, based upon the following questions:[6]

1.   **Does the tool provide the means for developing a disaster recovery plan for my entire company or is it limited to data processing and/or telecommunications recovery?** If disaster recovery planning is to be comprehensive, covering all vital corporate functions and assets, the planning tool selected must be able to handle plans for the recovery of more than hardware, software, and electronically stored data. Most planning tools do not provide this capability in their noncustomized form, despite vendor claims to the contrary. Thus, if the planner is seeking to use the PC-based planning tool to plan for total corporate recovery, this should be known in advance and specific questions should be asked about the degree to which the plan can be modified to meet customer needs.

---

[6]For additional information, refer to Jon William Toigo, "Disaster Recovery Planning Tools for the Microcomputer," *Journal of Accounting and EDP*, Vol. 4, No. 3 (Fall 1988): 43–49.

2. **Does the plan tool require that a particular methodology or approach to disaster recovery planning be adopted that differs from the methodologies used in other planning activities?** Good disaster recovery planning differs little from other types of business planning. Objectives are developed, tasks are derived from objectives, and criteria are set forth to gauge task and objective fulfillment. An experienced planner could use basic project management skills to develop and maintain an effective contingency plan. Novice planners may need more than a generic project management software package to develop their first plans, but the package that they use should not deviate drastically from a basic project management approach. If a manual is required just to understand the methodology that will be used with the PC-based plan, it is probably not the plan the user should purchase.

3. **Is the plan tool comprehensive?** This may be difficult for the novice planner to judge. At a minimum, the plan should offer sections in the following areas:

    - **Action Plan**—The order in which recovery activities must be undertaken to result in the speedy conclusion of a disaster situation.
    - **Plan Activities**—The tasks that must be undertaken in a recovery situation. These should be organized by some means (a numbering system, for example) and related to an Action Plan.
    - **Recovery Teams and Notification Directory**—The plan tool should have a location for recording the names of company personnel who will play a role in a recovery situation, as well as a list of telephone numbers for all personnel who must be notified in the event of a disaster.
    - **Vendor Information and Contact Directory**—The plan tool should provide a location for recording information about all vendors who will provide products or services in a disaster and the names and telephone numbers of vendor contacts.
    - **Records Requirements and Locations**—The plan should include a section detailing the location and type of vital records stored off-site and the procedure for obtaining them for use in recovery.
    - **Equipment Inventories**—An inventory of hardware and other equipment needed for recovery should be maintained in the plan, both for insurance purposes and for use as a checklist in plan testing.
    - **Communications Networks and Line/Equipment Requirements**—The plan should provide a description of network operations and recovery requirements in terms of lines and equipment or services.
    - **Application Systems Software and Hardware Requirements**—This plan section should provide not only system descriptions and prerequisite hardware for operations, but also user hardware requirements.
    - **Company Information**—Information about company lawyers, insurance policies, lines of credit, and so forth, should be maintained in the plan document.

4. **Is the package user-friendly?** An excellent disaster recovery planning application should be no less user-friendly than any other software package. In fact, given the specialized work of the package, the planning tool should be more user-friendly. Here are some of the factors that contribute to user-friendliness.

   - **User Interface**—The more complex the package, the less expert the user should need to be to use it. This rule of thumb applies to nearly all applications. A well-designed package, therefore, should provide menus, ergonomic displays, and straightforward escape routes. Mouse controls might also be a plus, though this would reduce system portability if a fallback to cursor control were unavailable.
   - **Help Screens**—Given the subject matter, generous use of contextual help screens would be a definite advantage. A printed manual is no substitute for a help screen that may be invoked from anywhere in the program explaining to the user what he or she is about to do (or has done) wrong and what the options are.
   - **Tutorials or Samples**—One of the primary justifications for purchasing a disaster recovery planning tool is to get an idea of what an actual plan looks like. The PC-based tool should come equipped with a sample that can be modified by the user to accommodate his or her company requirements.
   - **Input**—Inputting data into the plan should be as simplified a procedure as possible. This issue is key to the debate over which type of planning package—word processing-driven or database-driven—is the best. Structurally, the plan should be organized to reduce or eliminate redundant data entry. It may be useful if the plan tool allows the importation of outside files through batch conversion or some other method since some of the information needed for the plan may have been assembled in another form or for another purpose using another system.
   - **Output**—Output from the planning tool should take the form of an actual plan. It should be divided into sections by subject or task and should be print-definable to accommodate documentation standards employed by many companies.

5. **What is the pedigree of the planning tool?** Many disaster recovery planning tools were developed by consulting firms for the consultant's use at customer sites. In some cases, the package was subsequently licensed for use by the customer who maintained the plan that the consultant created. Consulting firms began to customize their planning tools for use by disaster recovery backup facility vendors and their clients. Finally, direct sales to the customer became a lucrative source of business for consulting firms. Hence, many planning tools have a history of having been used to develop actual plans for real companies. The purchase of an untested planning tool may have the disadvantage of producing inadequate plans, a fact discovered only after testing or—in the worst case—in an actual disaster. Thus, the pedigree of the plan is extremely important.

DISASTER RECOVERY PLANNING TOOLS EVALUATION CHECKLIST

Product Identification

Product Name: _____

Vendor: _____

_____

_____

Scope	YES	NO	
Data Center Only			
Company wide			
Corporation wide			

Methodology	YES	NO	
Project Management (If NO, state other)			

Plan Description *(YES, if feature is provided)*	YES	NO	
Generic Plan			
Action Plan			
Plan Activities			
Recovery Team Directory			
Vendor Information			
Equipment Inventories			
Records/Locations			
Network Descriptions			
System Descriptions			
Company Information			

User-friendliness *(YES, if feature is provided)*	YES	NO	
User Interface (Menus, windows, mouse, etc.)			
Help screens (Contextual)			
Input Methods (non-redundant data entry, batch mode)			
Output Methods (diversity of reports, query language)			

Price:

Figure 3.5 Disaster recovery planning tools evaluation checklist.

These are several rudimentary considerations that will play a role in the selection of the best disaster recovery planning tool for a given company, if one is to be selected at all. In addition to these, the price of the planning tool, the availability of telephone support, and other factors that go into the selection of any software package will also be important.

However, disaster recovery planning tools are not necessary for the development of a disaster recovery plan. Modern office automation applications (work processors, spreadsheets, databases) from Microsoft Corporation, Novell Corporation, and others provide integration capabilities sufficient to develop a maintainable plan. Moreover, the adoption of the wrong disaster recovery planning tool may lead to serious delays in plan development while the planner learns both the package and its underlying methodology.

Other Control Vehicles

In addition to documentation and software standards, establishing other procedures in advance will help to circumvent potential problems that could delay the project. These procedures fall under three categories: team meetings, periodic reporting requirements, and problem intervention.

Team Meetings

It should be made clear to team members that attending planning meetings, which may release them from other work, is a business function. Formal starting and stopping times should be announced and observed and a formal agenda developed. Attendance at team meetings is very important. A procedure should be set up for notification in the event that a member must be absent.

Managers of team members must also be apprised of team activities and meetings. If a manager cannot make a team member available for the amount of time required to attend team meetings and to perform assigned disaster recovery planning tasks, this must be known so that a replacement team member can be found.

Team meetings provide the planner's only means for monitoring task fulfillment. They should be run professionally as forums for evaluating work and scheduling new tasks. Meetings are also an opportunity for team members to vent opinions and discuss options. They provide a means for the plan coordinator to offer advice or commend effort.

Periodic Reporting

The development of periodic progress reports should be asked of each team member. Reports should be in a word processing format so they may be readily integrated with periodic reports to senior management. Obtaining progress reports in advance of a team meeting will facilitate the development of an agenda to resolve problems.

Intervention

There are two areas in which intervention by the plan coordinator may be needed. The first is in surmounting the reluctance of a department manager to participate in data collection efforts or to make available a staff member to participate on the planning team. The second type of intervention may involve non-performing team members.

Resolving management indifference to the planning project may require a meeting between the project leader and the department head. During this meeting, it should be clearly explained to the manager that the project that is being undertaken is for the good of the whole company and that his or her participation would not only be appreciated but would guarantee that the plan will incorporate the needs of the subject department. If all else fails, the planner may need to seek the assistance of senior management.

The nonperforming team member poses another potential obstacle to plan fulfillment. The project coordinator should meet with the team member and a third party, such as the project secretary, a representative of the team member's department management, or a representative from the company human resources department to discuss the performance problem. Remedies depend on the attitude and reasoning of the team member. If the team member has too much regular work or would simply prefer not to be involved, a replacement should be sought. If the team member does not understand the tasks that he or she has been assigned or finds them difficult to perform, the planner should work with the team member to surmount the difficulty.

DEVELOPING THE PLANNING BUDGET

Developing a planning budget is not a complicated task. Projected time requirements will be available following the first meeting of the team. All that will be required is to identify resource requirements and billable items for the analysis phase. (Implementation and Evaluation Phases are budgeted later.)

Fixed Costs

If an office is established for the disaster recovery planner, furniture, power, computer equipment, and other fixed costs will comprise a line item on the analysis phase budget. Telephone bills may be estimated for inclusion into this figure depending on company policy.

Payroll Allocation

Ideally, the disaster recovery planning project will have at least two full-time personnel: the planner and the project secretary. Additionally, team members, "borrowed" on a part-time basis from other areas, may require a billing code to which they may bill hours spent performing work on behalf of the project. In some cases, the cost for their time is absorbed by their own departments.

If the project coordinator will be responsible for verifying time spent on project activities and will need to obtain payroll funding for planning team participants, it will be necessary to estimate for budget purposes the amount of time the project will be utilizing each team member's services. Time estimates will need to be translated into dollar costs for payroll using each team member's salary. The total of these payroll costs will be added to the planner and project secretary salaries to provide a payroll entry on the budget.

Travel

Expenses may be accrued for travel if the planner or team members will be visiting vendor disaster recovery facilities or attending conferences during the analysis phase. These should be estimated and included in the project budget.

Consulting Fees

If an outside consultant will be used during the risk analysis phase, anticipated costs should be listed on the budget. In some cases, consulting services will need special approval and will be obtained through management rather than billed to the disaster recovery project directly.

Supplies

Office supplies, computer disposables, and other materials may be required for the project during the analysis phase. These should be anticipated and accounted for in the budget to the greatest possible extent.

Other Items

The acquisition of special software for risk analysis or disaster recovery planning tools may be classified under the catch-all category of other items.

SUMMARY

In this chapter we have examined the typical initiation steps for a disaster recovery planning project. These have included conducting an informal risk assessment, presenting to management on the need for planning, assembling a project team, developing the team mission and objectives, instituting project controls, and establishing a budget for the next phase of the project. Having accomplished these steps, the planner is ready to proceed with the first phase of disaster recovery planning: data collection and risk analysis.

Checklist for Chapter

1. Using industry statistics, case histories, legal and regulatory data, prepare a risk assessment.

2. Using the risk assessment, draft a memo underscoring the need for corporate-wide disaster recovery planning.

3. Conduct a management awareness briefing to seek senior management's support and approval to conduct formal risk analysis.

4. Draft a corporate policy statement on disaster recovery planning that would be suitable for your company and management.

5. Select a project team and justify your selections on the basis of expertise and background of identified personnel and organizational factors.

6. Evaluate the benefits and drawbacks of hiring a consultant to aid in plan development. Identify specific areas of the plan where you believe a consultant may be of assistance.

7. Develop an action plan for the first meeting of the disaster recovery team. The plan should include:

   - A mission statement or general objective for the project.
   - A list of project objectives, tasks, and deliverables.
   - Draft data collection tools and an overview of how the tools will be administered and used in risk analysis. (Consult Chapter 4 and 5)
   - Standards for documents and software.

8. Optional: Evaluate specialized disaster recovery software for applicability to your plan requirements.

9. Conduct a meeting of the project team. Explain the approach to be taken to disaster recovery planning. Schedule future meetings. Assign data collection tasks. Assign task leaders for future planning activities. Revise the data collection and survey forms. Identify system and network documentation requirements.

10. Prepare a budget for the analysis phase that includes the following:

    - Fixed costs
    - Payroll
    - Travel
    - Consulting Service Fees
    - Supplies
    - Other expenses

To obtain information on potential natural disaster risks in your geographical region, write to: FEMA, Office of Risk Assessment or Office of Natural and Technological Hazards, 500 C Street SW, Washington, DC 20472; National Hurricane Center, 1320 South Dixie Highway, Room 631, Coral Gables, FL 33146; National Severe Storms Forecasting Center, 601 East 12th Street, Room 1728, Federal Building, Kansas City, MO 64106.

Write to the NFPA at Batterymarch Park, Quincy, MA 02269.

Contact the National Institute for Standards and Technology, US Department of Commerce, Gaithersburg, MD 20899, and the Information Industry Association, 555 New Jersey Avenue NW, Suite 800, Washington, DC 20001, for information on private and public computer security studies.

Even in the case of governmental computing, where regulatory requirements are quite exact, there seems to be little enforcement and only the smallest of budgets for developing and for auditing disaster recovery plans.

This matrix was originally published in Jon William Toigo, "The Disaster Recovery Plan: Who Should Develop It?" Portfolio 85-01-30, *Data Security Management* (New York: Auerbach Publishers, 1988): pg. 11.

For additional information, refer to Jon William Toigo, "Disaster Recovery Planning Tools for the Microcomputer" *Journal of Accounting and EDP*, Volume 4, Number 3, Fall 1988, pp. 43-49.

4

Data Collection

OVERVIEW AND METHOD

Once management sponsorship has been received and the disaster recovery project has been formally initiated, the next major task to be accomplished is data collection. Data about business functions and their resource requirements are needed to perform risk analysis, a process that seeks to put a price tag on downtime by identifying the dollar costs associated with a business function interruption of a specified duration.

As depicted in Figure 4.1, the disaster recovery coordinator and project team need to explore the business functions that are provided by each department, division, or work unit within the company. They must break each function down into its component tasks to analyze its system, network, personnel, and logistical requirements. Only through this *decomposition* analysis, can the relative importance and relative vulnerability of each business function be assessed.

Thus, the data collected at this point in the disaster recovery project serves several purposes.

- It is the basis for estimating loss potentials and ultimately for cost-justifying disaster recovery planning to senior management.
- It is used to identify cost-effective applications for disaster prevention products.
- It provides a clear definition of data and records storage requirements.
- It helps to set objectives and parameters on strategies that will be developed to recover business functions in the event of disaster.

THE DISASTER RECOVERY PROJECT
DATA COLLECTION

MAJOR ACTIVITIES

Administer Business Function Survey
Questionnaires to End User Work Units

Create Database of Business Functions

Assess Outage Impact Costs

Perform Criticality Assessment

Identify Automation Resource Dependencies

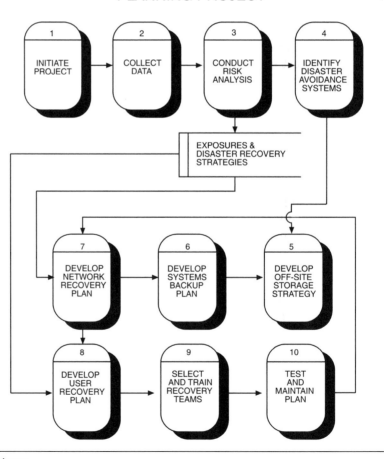

Figure 4.1 The data collection phase.

The following is a generic procedure for undertaking data collection. It involves the administration of a multipage questionnaire (with supporting worksheets) and the collection of quantitative data on automation resources used to support business functions. Typically, questionnaires are followed up with personal interviews between disaster recovery project staff and business unit managers or senior staff to develop a comprehensive understanding of business functions and their resource requirements.

The methods and forms depicted here have been employed successfully in the disaster recovery planning projects for several medium-sized financial companies. With minor modifications, there is no reason why they could not be adapted to support data collection in larger or smaller businesses or in different industry areas.

About the Questionnaire

Figure 4.2 provides the full business function survey instrument to be employed in a generic data collection effort. Seven pages in length, the questionnaire is used to record several types of data.

Respondent Identification Data

The top portion of the survey instrument identifies the business unit that is the object of the survey and the manager or senior staffer from the business unit who has prepared the response. Also recorded are the date of the survey, the number of personnel in the business unit, and the physical location of the business unit.

Business Function Description

The first formal section of the instrument is used to describe, in a chronologically ordered listing, the business functions of the work unit. For example, the loan processing division of a mortgage banking company may have the following functions:

1. Take a customer application for a loan over the telephone, prequalifying the buyer, and setting up a loan application package mailout.

2. Receive the completed loan application and set up the loan on a processing system.

3. Process the loan application.

4. Review the loan application prior to submission to Underwriting Department.

5. Submit the loan to Underwriting Department.

6. Handle customer inquiries about loan status.

7. Notify the customer of the outcome of customer review and submit processing file to Closing Department.

CONTINGENCY PLANNING PROJECT
BUSINESS FUNCTION SURVEY

INTERVIEWEE:	DATE:
TITLE:	**INSTRUCTIONS:** Please complete the following
DEPARTMENT/DIVISION:	survey as completely as possible. Attach additional
	sheets as necessary. Return completed survey to Jon
NUMBER OF PERSONNEL IN DEPT/DIV:	Toigo, Disaster Recovery Coordinator, Room 123, MIS
PHYSICAL LOCATION:	Division. Thank you.

SECTION ONE: BUSINESS FUNCTION DESCRIPTION

INSTRUCTIONS: Please identify the business functions performed by your department, division or work group. If the business functions performed have some sort of chronological order (e.g., work "A" is performed before work "B"), please indicate this relationship by numbering functions in order (i.e., "1. Take Credit Application. 2. Process App.").

ARE THE BUSINESS FUNCTIONS SHOWN SEQUENTIAL OR CHRONOLOGICAL? (Circle One)

Yes   No

PAGE 1 -- Attach additional pages as needed.

Figure 4.2 Business function survey.

SECTION TWO: WORK BREAKDOWN

INSTRUCTIONS: For each business function performed by your department, division or work unit in Section One, please provide a breakdown of major tasks. If the business function being described is "1", label component tasks "1a", "1b", etc. Be sure to indicate system or network usage in each task. Attach additional pages as needed.

ARE DEPARTMENT EMPLOYEE MANUALS AVAILABLE TO DOCUMENT DESCRIBED WORK?   (Circle One)

Yes   *No*   (If YES, please attach manuals for reference.)

PAGE 2 -- Attach additional pages as needed.

Figure 4.2 Business function survey. (Continued)

SECTION THREE: RESOURCE REQUIREMENTS

INSTRUCTIONS: For each business function performed by your department, division or work unit in Section One, please provide a list of resources required to perform each function. The list should include pre-printed forms, office equipment (i.e., photocopy machines, postage meters, etc.), computer devices (terminals, PCs, printers, etc.), telecommunications and network capabilities (modems, FAX machines, telephones, WATS lines, telephone messaging/announcing equipment, etc.), and records (hardcopy, microform, electronic databases). If appropriate, associate resource requirements with specific tasks identified in SECTION TWO. Attach additional pages as needed

FOR COMPUTER AND TELECOMMUNICATIONS EQUIPMENT, HAVE YOU COMPLETED A TERMINAL USAGE PROFILE FOR EACH RESOURCE? (Circle One)   *Yes*   *No*   (If YES, please attach)

PAGE 3 -- Attach additional pages as needed.

Figure 4.2 Business function survey. (Continued)

SECTION FOUR: OUTAGE IMPACT ESTIMATE

INSTRUCTIONS: The following series of questions relate to the financial impact of a disaster that interrupts the normal work of your department, division or work unit. Using all available data, you are asked to estimate the cost to the company of a 24-, 48-, and 72-hour interruption in the business functions performed by your organization. There are no right or wrong answers. You will need to explain your responses, especially when you assign dollar costs to such non-quantifiable items as ''lost sales'' or ''lost client confidence''.

OUTAGE IMPACT BY BUSINESS FUNCTION (Photocopy this page and complete for each business function identified in SECTION TWO.)

FUNCTION: _____

A **24-hour** interruption of this function would be: (Check one)

A. Disastrous and non-recoverable
B. Recoverable at high cost (i.e., would require substantial catch-up work)
C. Relatively insignificant

A **48-hour** interruption of this function would be: (Check one)

A. Disastrous and non-recoverable
B. Recoverable at high cost (i.e., would require substantial catch-up work)
C. Relatively insignificant

A **72-hour** interruption of this function would be: (Check one)

A. Disastrous and non-recoverable
B. Recoverable at high cost (i.e., would require substantial catch-up work)
C. Relatively insignificant

(If you checked A or B in any of the above, explain your answer on a separate sheet. On the same page, estimate the costs of an outage and explain your estimate. Be sure to label each explanation sheet with the name/number of the subject business function.)

Are there any particular times of the month in which an interruption would be especially costly?

No. This function is not calendar-sensitive.
Yes. (Explain below.)

_____
_____
_____
_____

PAGE 4 -- Attach additional pages as needed.

Figure 4.2 Business function survey. (Continued)

SECTION FIVE: EXISTING CONTINGENCY PLANS

OVERVIEW: The following series of questions cover preparations you may have already made for coping with the possibility of an interruption. Such preparatory measures may include policies and procedures for removing copies of key documents or data off-site, schedules for backing up PC hard disks, requirements for documenting   spreadsheets, databases or applications developed within your department, etc. Also of interest are any installed disaster prevention systems (i.e., sprinkler systems, fire extinguishers, etc.) and any activities that your work unit may undertake to familiarize staff with building evacuation plans and other emergency procedures.

Does your work unit utilize local computing (i.e., PCs, local area networks, minicomputers)? (Circle one)

$\qquad$ *Yes* $\qquad$ *No*

If YES, are backups of these systems and the data they contain made regularly? (Circle one)

$\qquad$ *Yes* $\qquad$ *No* $\qquad$ *Backups performed by another department*

If BACKUPS are performed by another department, identify the department: _____

Are data or records produced by your work unit required for day-to-day operation? (Circle one)

$\qquad$ *Yes* $\qquad$ *No*

Are data or records produced by your work unit subject to legal or regulatory retention requirements? (Circle one)

$\qquad$ *Yes* $\qquad$ *No*

If you circled YES in either of the two preceding questions, identify the method used to retain records: (Check one)

A. Copies or originals stored off-site
B. Copes or originals stored in fire-proof cabinets or safes
C. Copies or originals retained in files on-site
D. Other (Identify: _____)

What disaster prevention capabilities are installed in or near your work area? (Check all that apply)

A. Fire supression systems (Chemical or sprinklers)
B. Fire extinguisher units or hydrants
C. Smoke detectors
D. Access controls (combination locks, card key units, etc.)
E. Uninterruptible power supplies or surge suppressors
F. Other (Identify: _____)

Continue to next page.

Figure 4.2 Business function survey. (Continued)

SECTION FIVE: CONTINUED

Does your work unit utilize spreadsheets, databases or other programs developed by in-house staff? (Circle one)

Yes　　　　　　*No*

If YES, are these programs vital for work performed by the unit? (Circle one)

Yes　　　　　　*No*

If YES, are custom programs documented both in terms of design and user procedures? (Circle one.)

Fully documented　　　*Design only*　　*User Procedures only*　　　*Not documented*

If DOCUMENTED, are copies of the documentation stored off-site or in fire-rated cabinets or safes? (Circle one)

Yes　　　　　　*No*

Is there a formal program of emergency awareness training in your work unit?

Yes　　　　　　*No*

If YES, does it cover: (Check all that apply)

☐	A. Fire escape routes
☐	B. Detection/suppression equipment operation
☐	C. Equipment/records evacuation
☐	D. Computer security
☐	E. Emergency notification (Who to notify and when)
☐	F. Other (Identify: _____)

If you have a training and awareness program, how freqently is it conducted? (Check one)

☐	A. Monthly
☐	B. Bi-monthly
☐	C. Quarterly
☐	D. Annually
☐	E. As needed
☐	F. Other (Identify: _____)

Continue to next page.

Figure 4.2 Business function survey. (Continued)

SECTION FIVE: CONTINUED

To your knowledge, is your work unit covered by business interruption insurance? (Circle one)

Yes   No   Unknown

If known, identify the type of coverages provided, the insurance carrier and the annual premiums paid:

_____

_____

_____

_____

_____

Does your work unit have a formal plan for coping with an emergency outage or disaster? (Circle one)

Yes   No

If YES, please provide a copy of your plan.

In the event of a disaster, could your business functions be provided by a smaller staff for a limited period of time?

Yes   No

If YES, how many staff members would be required at an absolute minimum? Explain your response.

_____

_____

_____

_____

_____

THANK YOU FOR YOUR ASSISTANCE!

Please direct your responses to: Jon Toigo, Disaster Recovery Coordinator, Room 123, MIS Division.

A representative of the Disaster Recovery Planning Project will be in contact in the near future to schedule a follow-up interview, if necessary.

Figure 4.2 Business function survey. (Continued)

These seven steps occur in a more-or-less orderly process and are interdependent in the sense that function two can be completed only after function one, and so on. This sequence is reflected in the order in which the steps are listed and numbered. In the case of nonsequential or nonchronological business functions, such as the functions of a company human resource department, functions may be numbered in any order that is convenient. A question on page two of the questionnaire provides the respondent with an opportunity to identify whether business functions are chronological.

Work Breakdown

The next section of the questionnaire seeks to break down each of the previously listed business functions into their component tasks. In this way, resource requirements can be more readily identified. For example, the first business function in the case of the mortgage bank loan processing division previously cited was summarized in a simple statement: "1. Take a customer application for a loan over the telephone, prequalifying the buyer, and setting up a loan application package mailout." In the Work Breakdown section of the survey, the respondent lists, using an a-b-c format, the specific tasks involved in the function. For example:

1a.   Receive customer telephone call over toll-free line.

1b.   Route call to first available loan counselor.

1c.   Discuss customer needs and compare to available loan programs.

1d.   Prequalify buyer using loan prequalification program.

1e.   If buyer prequalifies, take address information and type mailing label for customer loan application package.

1f.   Affix label to preassembled loan application package and place in Mail Out basket.

It is important to solicit employee manuals or other documentation to support the work breakdown that is provided. A question at the end of the section requests that documentation, if available, be attached with the questionnaire response.

Resource Requirements

Using the business functions enumerated in the first section of the survey, supported by the work breakdown in the following section, the next part of the questionnaire asks the respondent to identify resources required to provide each listed business function. Resource requirements include computer systems, telecommunications systems, office equipment, preprinted forms, records, and any other special supplies that need to be stated. Two additional forms are provided to aid in this investigation.

First, as shown in Figure 4.3, a Resource Requirements Worksheet may be used to facilitate the respondent's efforts. This worksheet provides a categorical, question-and-answer format for recording resource requirements data. (This is in contrast to the blank-page format in the master questionnaire.) Thus, the worksheet may be very useful in obtaining comprehensive information, including requirements that may have been overlooked in a free-form respondent description.

RESOURCE REQUIREMENTS WORKSHEET

BUSINESS FUNCTION: _____

TASK: _____

NUMBER OF PERSONNEL INVOLVED: __   **DURATION OF TASK:** _____

PEAK TIMES: _____

AUTOMATION REQUIREMENTS:

☐ MAINFRAME TERMINALS (Specify Number: _____)
☐ MINICOMPUTER (Specify Type and Model: _____)
☐ LOCAL AREA NETWORK (Specify Type: _____)
☐ STAND-ALONE PC (Specify Type: _____)

WHAT SOFTWARE/SYSTEMS ARE USED?

_____
_____
_____
_____

PERIPHERALS? (i.e., PRINTERS, DISK STORAGE SUBSYSTEMS, MOUSE, etc.)

_____
_____
_____

TELECOMMUNICATIONS REQUIREMENTS:

☐ MODEM
☐ FAX
☐ VOICE SERVICES (Circle Applicable: Local   Long Distance   Inbound   Outbound)
☐ PROPRIETARY NETWORKS

OFFICE EQUIPMENT REQUIREMENTS:

☐ PHOTOCOPY EQUIPMENT
☐ MICROFORM READER
☐ CALCULATORS
☐ TYPEWRITERS
☐ OTHER (Describe: _____)

CONTINUE TO NEXT PAGE

PAGE 1 OF 3

Figure 4.3 Resource requirements worksheet.

SUPPLY AND LOGISTICS:

Are pre-printed forms required? *Yes   No*

If YES, list forms and supplier, and indicate whether forms are continuous or single sheet ("C" or "S").

FORM NO.	FORM NAME	SUPPLIER	CONTACT	ADDRESS/PHONE	C/S

OTHER SUPPLIES (List with supplier contact information. Attach more pages as needed.)

_____

_____

_____

_____

_____

_____

_____

_____

_____

_____

Figure 4.3 Resource requirements worksheet. (Continued)

RECORDS REQUIREMENTS:

Are historical records or data required? *Yes   No*

If YES, list required records/data and their format (P=paper, M=microform, E=electronic). For electronic records/data, specify source of data and name/location of electronic files. Attach sheets as needed.

RECORD/DATA DESCRIPTION	FORMAT P, M or E	FOR E-TYPE RECORDS/DATA

Figure 4.3 Resource requirements worksheet. (Continued)

A separate worksheet should be completed for each business function listed on the master questionnaire. Taking a closer look at this form, the following information is recorded:

- The business function and specific task(s) to which the worksheet pertains
- Number of personnel involved in the performance of the function under normal operations
- The duration of the task (how long the task takes to complete) under normal operations
- Peak times for the task (i.e., times of the day, week, month, or year that the task receives the most activity)
- Automation requirements for the task, broken down into hardware requirements, software/systems requirements, and peripheral requirements
- Telecommunications requirements
- Office equipment requirements
- Supply and logistics items requirements, such as pre-printed forms and other special supplies
- Records requirements

The Resource Requirements Worksheet thus supports the functional descriptions and work breakdown data in the main survey instrument to provide a comprehensive picture of what resources are required to provide a specified business function under normal operating conditions. The use of the worksheet also provides the means for obtaining raw data in a format that can be readily maintained in an electronic database should a planner elect to use such a database.

A second form, the Terminal Usage Profile shown in Figure 4.4, is used to describe the specific automation (system and network) resources used by company work units. Unlike the Resource Requirements Worksheet, data entered on the Profile form does not pertain to a specific business function, but to the device or terminal itself. Hence, this form is completed for each computer terminal, peripheral, or other automation resource used by the work unit in normal business operations. Data collected on this form includes:

- **Terminal Data**—The location (both physical and by department or work unit) of the device, the ID number and serial number of the device, plus controller and cable ID numbers.
- **Terminal Description**—The type of device, any connected peripherals, method of connection (in the case of a computer terminal, cabled to a controller or channel extender, or via modem), and special cabling requirements (information about Balun connectors, adapters, or special media requirements).
- **Usage Description**—A technical description of the applications with the terminal or other device is used, whether they are on-line or batch-request applications, and what the peak usage hours for the device are in a given work day.

TERMINAL USAGE PROFILE

TERMINAL DATA

DEPARTMENT: _____   LOCATION: _____

TERMINAL ID: _____   CONTROLLER: _____

SERIAL NUMBER: _____   CABLE ID: _____

TERMINAL DESCRIPTION IDENTIFICATION

Identify (1) Terminal Type, (2) Peripheral Devices, (3) Method of Connection, (4) Special Cabling Requirments

_____

_____

_____

_____

_____

_____

_____

_____

_____

_____

USAGE DESCRIPTION IDENTIFICATION

PRIMARY APPLICATION(S):

_____

_____

NUMBER OF USERS: ____   ON-LINE? ____   BATCH? ____

| PEAK USAGE HOURS: |
| *(In Military Time)* |
| FROM: _____ |
| TO: _____ |

COMMENTS: *(Note such details as peak monthly use, use of keyboard templates, data entry forms requirements, etc.)*

_____

_____

_____

_____

DATE: _____   COMPLETED BY:

Figure 4.4 Terminal usage profile.

As suggested in the previous description, much of the information collected on the Terminal Usage Profile form is technical in nature. Thus, this form may need to be administered by a technical member of the disaster recovery team on a case-by-case basis and completed using information maintained by corporate information systems or data processing departments.

The intent of the Profile is to describe each automation resource used by the company both technically and in terms of company work. It is not sufficient to know that five terminals are installed in the loan department. The planner must know whether the terminals are used constantly or rarely, whether they are used by individuals or shared by groups, whether they are used to request reports or to perform on-line functions, and so forth. This information, in turn, will be interpreted to design recovery strategies that will restore business functions with a minimum of scarce automation resources.

Outage Impact Estimate

With substantial data collected on business functions, their component tasks, and their resource requirements, the survey instrument turns its focus to the consequences of an interruption. What is sought are respondent estimates of the costs, in the absence of a disaster recovery plan, accruing to the interruption of normal business functions in their work units. (See page 4 of the Business Function Survey, Figure 4.2.)

Narrowly defined outage costs of a 24-, 48-, and 72-hour interruption are calculated by respondents using an Outage Estimate Worksheet, like that shown in Figure 4.5. Ideally, such cost estimates are prepared for each business function using any and all budgetary data (including "projected data" such as sales predictions based on historical performance) at the disposal of the manager. Then, returning to the survey instrument, respondents judge the impact of a 24-, 48-, and 72-hour interruption using qualitative criteria, classifying interruptions of varying lengths as disastrous, recoverable at high cost, or relatively insignificant. Since the timing of a disaster may have a direct bearing on its cost, the questionnaire also asks for a judgment regarding the calendar-sensitivity of the cost estimate (i.e., are there particular times of the month in which an interruption would be especially costly?). Respondents who indicate calendar-sensitivity are asked to explain their answers and to provide cost estimates for outages during peak work periods.

In the absence of reliable statistical data on the costs of business interruption, these outage impact assessments are about the best quantifications of disaster loss potentials that can be defined. Totalling the outage costs for all business functions, company-wide, can provide a compelling aggregate loss potential that may be used to cost-justify the expense of the disaster recovery project to senior management. These numbers may be used in other ways, during formal risk analysis, to "demonstrate" senior management's return on investment in disaster prevention and disaster recovery products and services.

In reality, of course, the impact estimates developed by work unit managers are *soft* numbers. Even when taken together, they cannot provide a true picture of disaster loss potentials confronting the company. Why? Respondents generally cannot predict or quantify the *collateral effects* of a disaster. That is, they generally have no means for estimating how their failure to provide business functions will impact other departments that utilize their output. This *ripple effect* is a multiplier of disaster costs that cannot be readily quantified. In reality, the total cost of a disaster is much greater than the sum of parts—the recovery costs of each work unit.

OUTAGE IMPACT ESTIMATE WORKSHEET

OVERVIEW: Following are some cost factors to consider when identifying the dollar cost of an interruption of critical data processing or communications services. This worksheet may be used to calculate the costs of a 24-, 48- or 72-hour interruption.

SAMPLE COST FACTOR	Cost after 24-hours	Cost after 48-hours	Cost after 72-hours
Loss of revenues based on average daily revenues			
Loss of orders or value of missed shipments			
Overtime for "catch-up" work			
Value of lost hardware			
Cost for records/data salvage			
Cost of restocking forms, supplies			
Lease costs			
Loss of customer confidence			
Potential legal costs			
TOTAL			

Figure 4.5 Outage impact estimate worksheet.

Still, attempts to quantify dollar loss potentials by work unit resulting from an interruption in the conditions required for normal operations have merit. These numbers are the grist of risk analysis and cost-justification and have even greater sway with management when they are known to have been derived from operational budgets and other data by work unit managers themselves.

It is worth noting that the Outage Impact Estimate Worksheet contains only nine categories of prospective cost factors:

- Loss of revenues based on average daily revenues
- Loss of orders or value of missed shipments
- Overtime payroll for catch-up work
- Value of lost hardware (possibly a depreciated loss)
- Costs for salvaging records and data
- Costs for restocking forms and supplies
- Cost for leasing a new or temporary facility
- Value of lost customer confidence
- Potential legal costs

These are only example categories and may be modified in content or number based on the needs of the particular company. Some costs may overlap, such as the loss of orders and the loss of customer or client confidence: The dollar value of the former may be the proper expression of the latter. How the cost factors are framed is relatively unimportant, provided that the factors identified by the respondent are cogent and their assigned values defensible by available operational data.

Existing Contingency Plans

The final three pages of the Business Function Survey consist of a series of Yes/No and multiple choice questions designed to uncover existing disaster prevention or preparation programs that may have been established within work units over time. Chances are very good that the possibility of disaster has occurred to someone at some time. If there has been any activity based on this concern, it might have resulted in programs, procedures, or policies that can be co-opted into the business-wide contingency plan.

The major subsets of this investigation into existing emergency plans include checks for:

- Data backup schedules for departmental computer systems (PCs, LANs, or minicomputers)
- Ongoing procedures providing for the retention and safe storage of critical data or records
- Installed disaster prevention technologies
- User and system documentation for custom application software, spreadsheets, and databases designed or developed by in-house personnel
- Ongoing programs of employee disaster awareness
- Business insurance policies and coverages
- The existence of any formal plans within the work unit for reacting to and/or recovering from a disaster

Minimum Staff Requirements

The final subject covered in the questionnaire, and one deliberately placed at the conclusion of the document, concerns work unit staffing requirements in a disaster recovery situation. While formal decisions about business function restoration priorities and strategies (including staffing requirements) for recovering from a disaster will not be made at this point, it is interesting to know how many personnel the manager believes would be needed to fulfill key roles and sustain minimum service levels in an emergency.

This question could be interpreted as inflammatory if placed at the outset of the questionnaire—possibly viewed as intimating an inefficient or overly large operational staff size. Placed in the resource requirements section, the question of resizing staff might blur other resource requirement considerations. Thus, this question needs to be addressed after the respondent has engaged in an exhaustive data collection effort and is thoroughly entrenched in the disaster recovery mindset. In follow-up interviews, it should be made clear that the objective of the question is to seek emergency staffing requirements, not to question normal operating requirements.

Other Data Collection Considerations

Of course, successful data collection involves more than the administration of a carefully designed questionnaire. A questionnaire, such as the one depicted here, is a tool for initiating interpersonal communication. The questionnaire opens discussion and starts interviewees to begin thinking about potentials for disaster and their consequences. The opinions of the interviewees, and their insights based on experience with the day-to-day operations of their departments or work units are just as valuable as quantitative data on the number of terminals installed in the office or the number of telephone calls handled through their work area daily. Raw data must be interpreted, and who is better than managers or staff—those who are intimately involved with the work—to make the interpretation?

This point cannot be made too strongly: for the data collection effort to result in the kinds of data required for effective risk analysis—and, ultimately, for the sake of plan integrity and solvency—the cooperation of business unit management and personnel is essential. This suggests a role for the disaster recovery coordinator that goes beyond mere questionnaire administration. The coordinator needs to be prepared to intervene to surmount the problems of unit manager indifference, their unfavorable task prioritization, or open hostility that may compromise cooperative efforts. Here, "carrot and stick" techniques may apply.

Recalcitrant business managers need to be encouraged to cooperate and participate in data collection. The "carrots" for eliciting this response include the following:

- **Team Players**—Explain to work unit managers that their participation is needed to ensure that the plan provides recoverability for the entire company: As members of the company management team, their responsibilities include aiding in such an important company-wide undertaking.

Conversely, explain that the failure of the manager to participate may have an impact on the solvency of the plan for recovering the other key business functions of the company in the wake of disaster. If possible, identify other managers who are providing their full cooperation to cajole the reluctant manager into participation.

- **Proactive Management**—Explain to work unit managers that their participation (or failure to participate) in the disaster recovery project reflects upon their perceived skill as managers. A good manager looks beyond day-to-day issues to plan for the development and safeguard the operation of the division or department. Anticipating potential interruptions and planning proactively to prevent disasters or to contain their harmful consequences are basic to good management.

- **Enlightened Self-Interest**—Some managers can also be persuaded to participate on the basis of their own self-interest. If it can be demonstrated that nonparticipation or delayed participation may result in the allocation of scarce disaster prevention and recovery resources to the work units of other more agreeable participants, the manager's self-interest may be enlisted to support the case for cooperation. Similarly, if it can be said that senior management is being apprised of manager participation and that this recognition has some positive value connected with it, the result may again be a more energetic and enthusiastic participation by the reluctant manager.

The obvious "stick" technique that may be brought to bear is implied in the self-interest "carrot." Senior management may need to be advised of the work unit manager's failure to cooperate with the data collection effort. This may result in a directive from senior management or even a censure of the non-cooperative manager—neither of which is particularly desirable in the eyes of most middle managers.

Of course, this "stick" should be used only after all other avenues to resolve the problem have been exhausted and only in cases where the manager's failure to cooperate cannot be rectified by other means. If data can be derived through other channels, such as procedures manuals and regular budgetary data, the disaster recovery team should seek these solutions and get on with their work. There is no point to cultivating hostility to the plan.

Moreover, disaster recovery team members, including the project manager, need to keep in mind the difficulty of the effort they are asking of managers. Completing the questionnaires and related documents previously described is a time-consuming task. Follow-up interviews promise to consume even more of the manager's precious time. Approaching the problem of noncooperation from this perspective will provide the data collector with a more facilitative outlook. What are needed are sympathetic, well-intentioned researchers, not emphatic, uncompromising missionaries.

OUTPUTS OF DATA COLLECTION ON BUSINESS FUNCTIONS

Following the administration of the business function survey instrument and related forms, the disaster recovery planning team will have amassed a large volume of data. Needless to say, an impressive stack of questionnaires hardly constitutes data in a refined form. To be useful, the data must be subjected to some basic analyses and synthesized into several information products. This section will examine some of these data collection outputs and their content.

Business Function Map

As previously indicated, the questionnaire method of data collection focuses on business functions. From the standpoint of disaster recovery planning, business functions are the objectives of recovery efforts. This is quite different from seeking to recover work units, systems, and networks to their predisaster form: often the first inclination of the novice planner and a frequent cause of planning failure.

Why seek to recover functions rather than organizations and support systems? The primary reason for the focus on functional, rather than organizational, recovery is economics. To recover an organization, planners must build strategies to provide work sites that are identical to disaster-stricken facilities. They must provide for identical resources—personnel, automation, records, and supplies. Finally, they must provide the means for a nearly instantaneous transition between sites in the event of disaster.

Such a strategy, known as *full redundancy*, is extremely expensive, requiring not only the construction and maintenance of identical facilities, but also ongoing data interconnections between the two sites. Typically, it is an option only for the largest corporations and for select government operations.

Moreover, even for those who can afford such a luxury, full redundancy is a myth. Accomplishing an instantaneous transition from the normal operating environment to the recovery environment using the same company personnel requires time. Personnel must be moved to the redundant site, which must be located remotely from the primary work site to prevent it from being affected by a regional disaster that impairs primary site operations. Assuming that no critical personnel have been injured, killed, or traumatized by the disaster itself, relocation plans can impacted by adverse weather or by the reluctance of personnel to separate themselves from their families in a time of crisis.

The probability of a completely debilitating disaster is extremely slight. In any disaster, however, there are restoration priorities. Coping with disasters that affect only portions of the company, rather than the whole company, under a full redundancy or organizational recovery strategy would be a much more time-consuming process than restoring just the functions of the work units affected. Proof of this assertion can be found in the experiences of companies that recovered successfully from severe disasters. Invariably, these cases demonstrate that functional recovery can be accomplished within 48 to 72 hours, while organization recovery—the restoration or transition to normal organizational models—may take several weeks following functional recovery.

For these reasons, business functions are the focus of data collection efforts. Once data has been collected on functions, their component tasks, and their resource requirements, a new map for the organization—a map drawn to indicate functional relationships rather than political or bureaucratic hierarchies—needs to be constructed. This is one of the information outputs of data collection.

Functional maps may be created using various diagrammatic techniques. Those familiar with structured systems design may find data-flow diagrams to be a useful method for depicting interrelationships between functions. In a data-flow diagram, arrows show data flows to and from function (or process) nodes, which are represented by bubbles or circles. Data sources are represented by rectangles, while stores or files are shown as straight lines. The most important feature of data-flow diagramming is its deemphasis of control in favor of data flow: producing a network representation rather than a control representation.

While this is not the place for an in-depth explanation of data-flow diagramming, it should be pointed out that a functional map of a business prepared in this method can reveal much about shared data inputs and outputs and redundant processes. Where the former exist, there may be an opportunity for consolidating logistical requirements. Where the latter, redundant processes, exist, there may be opportunities to reconfigure systems or networks to remove redundancies and provide necessary functions with less hardware, software, and resources. Tom DeMarco's *Structured Analysis and System Specification* (New York: Yourdon, Inc., 1978) remains the most readable book on data-flow diagramming and should be consulted by readers wishing to consider this technique for use in mapping the business functions of their companies.

A second diagrammatic technique, well-known to anyone who has ever sought to assemble a Christmas tricycle or troubleshoot a malfunction in a television or toaster oven, is flowcharting. A flowchart depicts the flow of control in a process or function. Unlike the uncluttered, simplified library of data-flow diagram symbols, the symbolism of flowcharting comprises a virtual language. There are symbols for processes, decisions, predefined processes, magnetic tape reels, magnetic disk storage, punched cards, documents, and much more. These symbols are connected by flow control arrows to provide a stream of decision-making logic undergirding the function(s) or process(es) being depicted.

The advantage of flowcharting, besides its relative familiarity among nontechnical persons, is that it can be used to depict procedural or sequential operations. In this way, it provides a shorthand method for showing chronological steps or tasks, separately or in relation to concurrent steps or tasks. In disaster recovery, where restoration priorities must be developed based on time and resource requirements, and on the input/output requirements of business functions, the ability to depict chronological events is very important. Data-flow diagrams show the flow of data without respect to chronology and are descriptive; flowcharts show the flow of control over data and are prescriptive or normative. Figure 4.6 depicts the difference in the techniques when used to model part of the work of the loan processing department described earlier.

Ideally, data-flow diagrams will be used to describe or map existing business functions so that their relationships and redundancies may be identified and alternative minimum system (and network) configurations can be developed and tested. Once these configurations have been developed, flowcharting will provide a useful technique for depicting the order of restoration priorities.

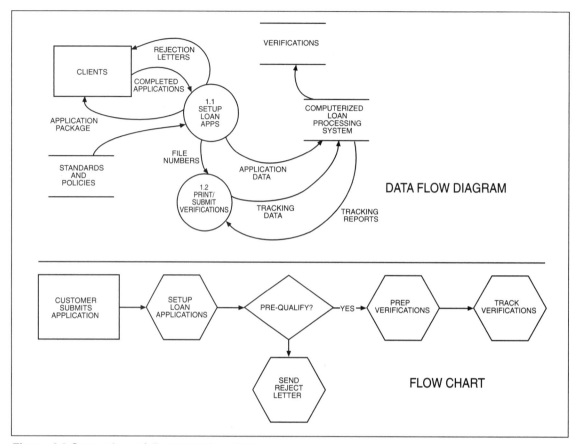

Figure 4.6 Comparison of diagrammatic techniques.

Organizing Collected Data

Whatever diagramming technique is used, the product should be a functional map for the company. The members of this functional map should be numbered or otherwise labeled to facilitate the creation of a series of files that will contain information about the inputs to the function, its outputs, and its resource requirements. Initially, each file will hold the completed questionnaire and any interview notes that pertain to the subject function. However, this file information will expand over the course of data analysis and as consultative data is acquired.

Figure 4.7 provides a file coversheet that lists some of the information a typical business function file will contain. In addition to the questionnaire, worksheets, and interview notes, the file will include:

- **Functional Map and Description**—The functional map for a specific business function is a close-up view or subset of the master business function map previously described. This subset will provide a description of the business function that is substantially more detailed than the overview map and should be accompanied by a descriptive narrative.

- **Criticality Assessment**—A criticality assessment is performed during risk analysis to identify which functions need to be restored immediately and which can be delayed. These restoration priorities are then integrated into strategies for systems, network, and user recovery that comprise the formal disaster recovery plan.
- **Disaster Prevention Worksheet**—Where possible and practical, the disaster recovery planning team seeks to minimize the possibility of disaster through the identification and installation of prevention technologies. An assessment will be conducted (see Chapter 6) to identify preventive measures that are appropriate to user work areas throughout the company and the results will be recorded on a worksheet that may correspond to a specific business function (i.e., when the function is presently provided by users in a common work area).
- **Notification Directory**—Notification directories are telephone listings of persons who should be contacted in connection to an interruption of the subject business function. The directory will contain the names of recovery team personnel (such as key users, managers, data processing personnel) and also of vendors who provide emergency supplies and services. This directory will be established after a recovery strategy has been developed for the subject function.
- **Inventories**—Various inventory lists may be prepared during recovery strategy development to identify resources that are to be provided at system and user recovery locations to facilitate recovery of the business function. Inventories may include magnetic media and records backups stored off-site, hardware and software products, supplies of pre-printed forms, office equipment, and so forth.
- **User Requirements**—A user requirements listing, also prepared during strategy development, identifies all requirements for transportation, lodging, and provision of work areas for users who are actively participating in recovery efforts.
- **Procedural Documentation**—This may include documentation collected at the time of work unit manager interviews detailing the work flow in the department or division, as well as documentation on custom-developed applications, spreadsheets, and databases implemented by work unit personnel. Other possible components include a backup schedule, records and data storage plan, and documentation on security and access control items such as program passwords, lock combinations, and so forth.
- **Recovery Priority and Procedure**—Once recovery strategy development is underway, the procedures to be observed by the recovery team will be documented. A copy should be maintained in the file to facilitate periodic review and update.
- **Change Management Forms**—Over time, work flows and their associated resource requirements may change, personnel turnover will occur, and new vendors will replace former vendors. All of these events will have an impact on the currency of the strategy for recovering the business function in the event of interruption. To maintain the plan, it will be audited periodically. These plan audits should produce change requests that will be logged and

documented in a formal change management system. Documentation of changes that pertain to the recovery of the subject business functions should be maintained in the file folder for that function.

BUSINESS FUNCTION FILE COVERSHEET

BUSINESS FUNCTION: _____

PURPOSE: _____

CONTENTS:

- [ ] QUESTIONNAIRE
- [ ] INTERVIEW NOTES
- [ ] FUNCTIONAL MAP AND DESCRIPTION
- [ ] RESOURCE REQUIREMENTS WORKSHEET
- [ ] IMPACT ANALYSIS WORKSHEET
- [ ] CRITICALITY ASSESSMENT
- [ ] DISASTER PREVENTION WORKSHEET
- [ ] NOTIFICATION DIRECTORY
- [ ] HARDWARE INVENTORY
- [ ] SOFTWARE INVENTORY
- [ ] RECORDS INVENTORY
- [ ] SUPPLIES/MATERIALS INVENTORY
- [ ] USER REQUIREMENTS
- [ ] PROCEDURAL DOCUMENTATION
- [ ] RECOVERY PRIORITY AND PROCEDURE
- [ ] CHANGE MANAGEMENT FORMS

Figure 4.7 Business function file coversheet.

Data Refinement

Once organized in this fashion, data about specific business functions is readily accessible for risk analysis and other uses. However, other basic data refinements can be performed at this time that will assist risk analysis efforts directly. These refinements are of two types.

First, data on specific business functions can be combined and sorted in ways that will provide revealing facts about loss exposures. Figure 4.8 provides one format for combined data. Using TOTALS data supplied in each Outage Impact Estimate Worksheet, a listing of business functions and their interruption costs can be assembled. Totals at the conclusion of this listing provide the aggregate outage costs accruing to a 24-, 48-, and 72-hour interruption. This data, in turn, may be used to plot a graph of outage costs over time. (Recall, however, that this data is *soft* and may be significantly less than actual costs, since cost multipliers are not reflected in the outage costs developed by questionnaire respondents.)

Another opportunity for data refinement is provided in Figure 4.9. This worksheet summarizes survey responses to questions regarding the impact of an outage of a specified duration (i.e., "Disastrous and nonrecoverable," "Recoverable at high cost," or "Relatively insignificant"). A preliminary criticality assessment may be demonstrated by grouping together those business functions for which a 24-hour interruption would be disastrous, recoverable, or insignificant, then repeating this categorization for the 48- and 72-hour time frames (an X is entered on the list for functions that confronted disaster at an earlier time period). Obviously, those functions that would confront disaster would be deemed "critical." Functions that would be recoverable only at high cost and those for which a 72-hour interruption would be insignificant may be classified as "vital" and "nonvital" respectively.

Of course, this assessment is very preliminary. It may turn out, upon review, that a function thought to be "nonvital" based on an outage impact cost estimate would create major costs within the context of functional interrelationships. For example, assume that an output of a "nonvital" function is needed for use as an input by a "critical" function. Without the input, the "critical" function cannot be performed. In such a case, the "nonvital" function must be regarded as "critical" and assigned a restoration priority that is equal to the "critical" function that uses its output.

Such dependencies will not be apparent from questionnaire responses, but they may be clear from the functional map and other documentation. Once the preliminary worksheet in Figure 4.9 has been completed, it would be worthwhile, time permitting, to reexamine the function diagrams of individual business functions as well as the map of all business functions to track inputs and outputs to each function and identify dependencies between functions. Comparing these dependencies to the worksheet may reveal that some functions need to be reclassified as vital or critical.

AGGREGATE OUTAGE IMPACT COSTS WORKSHEET

INSTRUCTIONS: Enter the TOTALS from OUTAGE IMPACT ESTIMATE WORKSHEETS for each business function. Attach additional pages as needed. Calculate total outage costs on page two, then graph totals for a 24-, 48-, and 72-hour interruption.

Funct No	Description of Business Function	24-hour outage cost	48-hour outage cost	72-hour outage cost

Page 1 of 2

Figure 4.8 Aggregate outage impact costs worksheet.

Funct No	Description of Business Function	24-hour outage cost	48-hour outage cost	72-hour outage cost
	TOTALS			

OUTAGE COSTS (Indicate scale)

	EVENT +24 HOURS	EVENT +48 HOURS	EVENT +72 HOURS

Figure 4.8 Aggregate outage impact costs worksheet. (Continued)

PRELIMINARY CRITICALITY ASSESSMENT WORKSHEET (EVENT +24 HOURS)

INSTRUCTIONS: Using Outage Impact Assessments from page four of the BUSINESS FUNCTION SURVEY, list all functions and their status as of EVENT +24 HOURS. List functions in the following order: first, all functions with "disastrous" outcomes, then all functions with "recoverable" outcomes, and finally, all functions with "insignificant" outcomes. Check the appropriate outcome column. Attach additional pages as needed.

Funct No	Description of Business Function	Disastrous impact	Recoverable at high cost	Impact insignificant

Figure 4.9 Preliminary criticality assessment worksheet.

PRELIMINARY CRITICALITY ASSESSMENT WORKSHEET (EVENT +48 HOURS)

INSTRUCTIONS: Using Outage Impact Assessments from page four of the BUSINESS FUNCTION SURVEY, list all functions and their status as of EVENT +48 HOURS. List functions in the following order: first, all functions with "disastrous" outcomes at EVENT +24 HOURS (place an "X" in the "Disastrous Impact" column, rather than a check mark), then all functions with "disastrous" outcomes as of EVENT +48 HOURS, then all functions with "recoverable" outcomes, and finally, all functions with "insignificant" outcomes. Check the appropriate outcome column. Attach additional pages as needed.

Funct No	Description of Business Function	Disastrous impact	Recoverable at high cost	Impact insignificant

Figure 4.9 Preliminary criticality assessment worksheet. (Continued)

PRELIMINARY CRITICALITY ASSESSMENT WORKSHEET (EVENT +72 HOURS)

INSTRUCTIONS: Using Outage Impact Assessments from page four of the BUSINESS FUNCTION SURVEY, list all functions and their status as of EVENT +72 HOURS. List functions in the following order: first, all functions with "disastrous" outcomes at EVENT +24 HOURS then at EVENT +48 HOURS (place an "X" in the "Disastrous Impact" column, rather than a check mark), then all functions with "disastrous" outcomes as of EVENT +72 HOURS, then all functions with "recoverable" outcomes, and finally, all functions with "insignificant" outcomes. Check the appropriate outcome column. Attach additional pages as needed.

Funct No	Description of Business Function	Disastrous impact	Recoverable at high cost	Impact insignificant

Figure 4.9 Preliminary criticality assessment worksheet. (Continued)

FUNCTIONAL INPUT/OUTPUT WORKSHEET

List data inputs to subject function by origin and content.

INPUTS

FUNCTION:

OUTPUTS

List data outputs from subject function by destination and content.

Figure 4.10 Functional input/output worksheet.

This interpretive or comparative refinement of raw data is the second type of "data refinement" mentioned at the outset of this section. To aid in identifying inputs to and outputs from a business function, Figure 4.10 is offered as a simplified worksheet. Using the *black box* method, the origins and content of inputs and the destinations and contents of outputs may be listed. The *black box* is the function itself. It is not necessary to complete this worksheet for every business function, if functional relationships are sufficiently clear from the map of the organization prepared earlier. However, they may be useful when troubleshooting a planning problem that appears to be rooted in a lack of relational clarity.

Based on the assessment of criticality, a list of critical, vital, and nonvital applications may be assembled. Figure 4.11 provides a format for the listing. Note that a code "O" or "R" should be entered for each listed function in the column labeled "BASIS." Enter "O" if criticality has been determined on the basis of outcome (i.e., the impact of an interruption), and enter "R" if assigned criticality is based on the relationship of the function to another function rather than the impact of the interruption on the function itself.

On the second page of the form, a Statistical Summary section provides the opportunity to record several statistical facts about the business functions that have been classified according to their criticality. Enter the total number of functions, followed by the total that have been classified as critical, vital, and nonvital. It may also be useful to record the number of records classified on the basis of outcome (O-BASIS) and on the basis of relationship (R-BASIS). A second column provides locations for expressing the number of listed functions by their classification as a percentage of total listed functions. These percentages may be useful during risk analysis in identifying dollar loss potentials represented by the interruption of critical, vital, and nonvital functions respectively.

This type of data refinement, based on the collating and sorting of survey responses, can be of great assistance in risk analysis. Of course, this type of refinement may be performed more expediently using database and/or spreadsheet applications on a microcomputer than by manual means. Whether automated tools will be used in data refinement is a decision for the project team. If electronic tools are employed, it is vital that applications be thoroughly documented and that data be frequently backed up and copies removed to secure storage. The disaster recovery planning team must set the example and adhere to the principles of disaster preparedness in all activities.

CRITICALITY ASSESSMENT LIST

INSTRUCTIONS: List all business functions by function number and description. In the CLASS field, indicate whether each function is CRITICAL (C), VITAL (V), or NON-VITAL (N) based upon outage impact assessments and relational factors. In the BASIS field, enter "O" if the assessment of criticality is being made on the basis of the outcome of an interruption on the subject function. Enter "R" if criticality has been assigned on the basis of the functional relationship between the subject function and another business function. Attach additional pages as necessary.

NO.	DESCRIPTION OF BUSINESS FUNCTION	CLASS C, V, N	BASIS O, R

Page 1 of 2

Figure 4.11 Criticality assessment list.

NO.	DESCRIPTION OF BUSINESS FUNCTION	CLASS C, V, N	BASIS O, R

STATISTICAL SUMMARY

_____ TOTAL # OF FUNCTIONS LISTED
_____ TOTAL # OF CRITICAL FUNCTIONS
_____ TOTAL # OF VITAL FUNCTIONS
_____ TOTAL # OF NON-VITAL FUNCTIONS
_____ TOTAL # OF "O" BASIS FUNCTIONS
_____ TOTAL # OF "R" BASIS FUNCTIONS

_____ PERCENTAGE OF CRITICAL FUNCTIONS (C-CLASS/TOTAL)

_____ PERCENTAGE OF VITAL FUNCTIONS (V-CLASS/TOTAL)

_____ PERCENTAGE OF NON-VITAL FUNCTIONS (N-CLASS/TOTAL)

Figure 4.11 Criticality assessment list. (Continued)

CONSULTATIVE DATA

In addition to the refinement of outage impact data and the derivation of loss exposure statistics, collective and interpretive methods may also be applied to such subjects as terminal usage and peak system (and network) utilization periods. However, it may be possible to obtain this information more readily from electronic usage logs and performance monitoring software that are often installed on corporate computers themselves.

System operation logs, telephone call accounting systems, and software-based performance monitors are all examples of technical data collectors employed for operational purposes that may be harnessed to provide important information within the context of disaster recovery planning. The information products of these data collectors may be useful in a number of ways.

- **Time-Sensitivity as a Determinant of Criticality**—There may be particular periods of the day, week, month, or year that the interruption of automation resources to a seemingly nonvital business function would create a disastrous outcome. To identify this potential, technical data collectors may be of use.

  Many automated system monitoring tools, for example, maintain access (log on) or job request records that identify system activity by user or by device (by terminal ID, for example). Often, sophisticated data sorting facilities are provided with the tool to assist in data analysis. Using these facilities and known terminal identifiers, it may be possible to identify periods of high volume usage—a clue to functional time-sensitivity.

- **Interruption Data**—In practically any system or network, there are occasional periods of downtime. If technical data collectors (or manual operating logs) record these instances, they may provide a microcosm for postulating the impact of an actual disaster.

  For example, if an application is taken off-line for several hours to perform hardware maintenance or to load a new software release, the volume of activity ("catch up" work) following restoration of the application to on-line status may provide insights into the work required to recover systems interrupted for a prolonged period by a disaster event. By closely examining post-interruption activity records, it may even be possible to observe the ripple effect—that is, how the temporary interruption of one application affects other applications in the same system.

- **Probability Statistics**—Everyone knows that disastrous interruptions may result from malfunctions in hardware or software, or from the failure of power, air conditioning, or other equipment support subsystems. What cannot be calculated effectively is the probability that such disaster potentials will be realized.

  Using problem history logs and technical data collector information, it may be possible to identify the frequency with which certain types of interruptions occur. From this incident information, according to some writers in the disaster recovery planning field, probabilities may be calculated that can be used in risk analysis formula. (See Chapter 5.)

The interpretation of data collected by technical means may require the skills of a corporate information systems (IS) or data processing (DP) staff member, if one is not already a member of the disaster recovery planning project team. This resource will also be needed to work with the company's automation professionals to acquire systems and network documentation that are needed for recovery strategy development.

Data Collection on Computer Systems

With the emphasis in the modern corporation on automation and technology, it is obvious that disaster recovery planning is, in large part, synonymous to planning for the recovery of automation resources that support business functions. Thus, in addition to the focus of the data collection effort upon business functions, a secondary thrust of this investigation must be directed to systems and networks themselves. What kinds of information are needed regarding corporate information systems?

First, general data—such as hardware and software inventories, configuration diagrams, and application descriptions—must be collected to orient the project team to the automation capability installed at the company. Extremely important is information about the data center itself, including support subsystems, facility layout and design, equipment locations and interconnections, physical access controls, and installed disaster prevention systems.

Working outward from this central point, information is needed regarding the dispersal and cabling of local terminal networks and also about connections—via data communications controller, private networks, or channel extension—to remote user populations. Also of interest are decentralized computing resources, including micro- and minicomputers and local area networks (LANs).

This may constitute a large volume of data, but for this information to be useful in the context of business function recovery planning, it must be reorganized according to functional resource requirements. The justification for this assertion is simple: Data processing systems and communications networks are not appropriate objects of disaster recovery planning. Just as disaster recovery planning does not seek to recover organizational structures in their original form, but instead an efficient, temporary organization designed to meet key functional requirements until a crisis has passed, so it is with automated systems and networks. The goal is to recover minimum system and network configurations that will support the specific requirements of business functions that have been identified as critical or vital.

To facilitate this effort, the disaster recovery planning team will need to know which applications (or facilities within applications) and their associated hardware platforms are directly related to specific business functions. Since some business functions are neither critical nor vital to the continuation of the business following a disaster, it stands to reason that efforts should be made to excise the automation resources associated with these functions from recovery system and network configurations. This minimalist approach, in theory, will save time and money, both of which may be in short supply over the course of plan development and in the stark reality of plan implementation.

To identify applications, systems, and networks associated with specific business functions, several information sources may need to be consulted. Questionnaire responses, for one, should identify automation resources used in conjunction with listed functions. However, the end user perspective is rarely comprehensive. Today's business systems, in a quest for user friendliness, often shield automation from the user. End users may believe that they are using different applications for different functions when, in fact, they are entering data into a common application with specialized data entry screens to provide the illusion of "one function, one application."

Moreover, end users typically have no concept of data storage methods (history files, key files, etc.) that are the stuff of applications developed in COBOL and other popular languages in business programming. They may not realize that they are working within the context of a large database (DB2 or IDMS) or that they are communicating with a mainframe through a front-end communications controller.

Thus, system documentation must be consulted to identify the actual configuration of the system in which transactions are taking place. With older systems that have evolved as a patchwork of enhancements and fixes, it may be difficult or impossible to separate automation resources along functional lines. Systems may need to be restored "as is" because they cannot be separated into functionally distinguishable entities without substantial reprogramming. Conversely, with newer database management systems (DBMS), it may be comparatively easy to distinguish units along functional lines—but, restoration of the entire system may be so resource- and time-efficient that it makes little sense to separate nonvital automation requirements from vital and critical resources.

Where, then, might a minimalist approach be worthwhile? It is in the area of decentralized systems that systems recovery becomes a many-headed hydra. The recovery of centralized systems is child's play compared to the difficulty involved in recovering many smaller systems, each running its own mixture of operating system, custom programs, and off-the-shelf business applications.

Often, custom applications running on departmental computers and local area networks have been developed without regard to programming standards, which are presumably the mainstay of mainframe programming departments, reflecting instead the peculiar preferences of their developers. In the realm of microcomputers, applications may be developed by inexperienced staff members with some computer savvy and a disdain for documentation.

What documentation exists of decentralized computer resources must be gathered and carefully reviewed and tested to see whether the purported designs of the applications are in fact the actual designs of the applications. This may be possible only through a review of source code and the performance of operational tests. For applications (databases or spreadsheets) developed using off-the-shelf software, a thorough audit of each application is a practical necessity. Standards for application testing and auditing, written by William Perry, are available through this publisher to aid the efforts of disaster recovery planning teams in validating systems documentation and creating such documentation where it does not presently exist.

What happens with the minimalist approach? Once decentralized systems are identified, their design and operation validated, and proper documentation assembled, planners must explore possibilities for *porting* decentralized applications associated with critical and vital functions to a centralized system. The rationale is a simple one: the greater the number of decentralized applications that can be moved to a single, centralized processor in an emergency situation, the fewer the number of CPUs that will need to be provided for in the plan, thus reducing implementation costs and recovery time requirements.

Portability rests on three assumptions. First, applications must be developed in a language for which a command interpreter is available on the destination CPU operating environment. Second, application software must be platform-independent in the areas of screen handling and data storage input and output (or adjustment must be able to be made readily to facilitate communication between the ported application and new devices and terminal types). Third, data files must be in a format compatible with the destination operating system.

The real issues of portability are, of course, more complex than just intimated. Current efforts on the part of IBM (so-called "program pipes"), the maturation of sequential query language (SQL), and the work of numerous groups to establish multiple operating system "partitions" within a single CPU hold promise as means for moving applications and data between platforms. These techniques need to be examined as an option to restoring automation resource requirements on multiple processing platforms—another task in the data collection effort.

Data Collection on Telecommunications Requirements

What has been said of data collection on systems applies also to voice and data communication networks, which are increasingly used to support business functions. Configuration diagrams and network operating descriptions must be obtained for key system-based or private branch exchange (PBX)-based networks that are typically employed to handle voice traffic both within the company and between the company and the outside world. Similarly, documentation is required on electronic data networks that connect the home office to geographically dispersed branch sites, or link the company to its clients and customers, or provide the means for data interchange between the company and financial institutions, regulatory agencies, or DP service vendors (i.e., remote payroll processing).

The first objective of this investigation is to learn what services communications networks provide at the company and, in technical terms, what facilities have been established to support these services. Additionally, researchers must acquire data about the non-company resources that facilitate company communication needs, including:

- The locations of local telephone company serving offices (COs) and long distance carrier points-of-presence (POPs) that control the switching of in-bound voice traffic to the company and provide gateways to wide area networks on out-bound calls (both current servers and possible alternates for these facilities that may be accessed in the event of a server outage).

- Special facilities, leased from public servers, such as T-1, satellite, or packet-switched radio networks, and backups for these services in the event of interruption.
- Line services, such as WATS, announcements, call queuing, and so forth, that are rented from telephone company servers, plus available services for on-demand rerouting of these services to alternative sites such as a backup data center or backroom operations center in the event of a disaster.
- Communications access facilities owned and maintained by building management in some non-company owned buildings.

Once this infrastructure for voice and data communications has been researched and documented, the next task is to attempt to associate facilities and services with specific business functions. As in the case of systems, end-user responses to the business function questionnaire both reveal and conceal what network resources are accessed and how they are used. End users often have no knowledge that dialing a long distance exchange causes the company PBX to route their call automatically to an outbound WATS trunk. They may not realize that the "current sales" report they are requesting from a computer menu is being generated from data obtained through a complex polling of computers at geographically remote branch offices, initiated by their request.

Network documentation may provide some of the behind-the-scenes information on these transactions. Other clues may be obtained from system (as opposed to network) documentation or by reviewing the output of call accounting systems, telephone dialing logs, message pad copies, or even telephone company statements.

Thus, a considerable effort may be required to identify the levels of network resource usage, and considerably more to identify the purpose of the usage in order to relate network facilities to business functions. There are no short cuts.

Fortunately, most company networks may be recovered intact in the event of an outage, mitigating to some degree the need for an exact accounting of which resources are used by which business functions. An exception is the company phone system. If emergency operations will be supported by a skeleton staff, the need for that 400-line PBX with all of its special user-programming features may not exist. Thus, data on the personnel complement of the recovery operations center will aid in sizing a replacement switch that may be more readily obtained and installed than original, full-featured equipment. Based on questionnaire responses, it may be possible to develop a recovery staff size estimate that may, in turn, be used to identify types and supply sources for emergency switching equipment.

Data Collection on Records Requirements

Another target for data collection is vital records and their retention requirements. Some records must be used in the day-to-day work of staff providing key business functions. Others must be securely stored and preserved, on-site or off, based upon legal or regulatory mandates. The former requirement should be adequately documented in the questionnaire. The latter may require additional research into legal requirements for records retention.

A number of professional associations have emerged over the past decade that focus on the informational and educational needs of records managers and records storage vendors, including the Association of Records Managers and Administrators (ARMA) in Prairie Village, Kansas, and the Association of Commercial Records Centers (ACRC) in Minneapolis, Minnesota. These associations, which may be locally represented, are an excellent source of data on records storage requirements and on other facets of records storage and salvage.

SUMMARY

Data collection is a challenging enterprise that does not end with risk analysis, but continues throughout the disaster recovery planning project. The data requirements identified in this section are a starting point for the collection effort. The targets and problems of collection may vary based on the company.

It may be useful for readers to review this book thoroughly and to orient themselves to the information requirements of disaster recovery planning project before they develop questionnaires and other strategies for data collection. The more comprehensive the information that can be collected at the outset of the project, the more expeditiously other project tasks may be undertaken and completed.

Checklist for Chapter

1. Review the model Business Function Survey instrument and tailor the form, and associated worksheets, to your business needs.

2. Obtain an organizational chart and list appropriate survey respondents.

3. Draft a memo stating the purpose of the survey, its authorization by senior management, and a deadline for response.

4. Direct the memo and a copy of the survey instrument to each name on the respondent list. Log questionnaires as they are returned and schedule follow-up interviews.

5. Schedule meetings with respondents who have not returned their questionnaires by the deadline date or who have otherwise indicated a refusal to cooperate with data collection efforts. Discuss their objections and seek to gain their participation. If cooperation is not forthcoming, seek other methods for obtaining needed data. If other methods are not forthcoming, seek the intervention of senior management.

6. Based on questionnaire responses, prepare a business function map for your organization using data-flow diagramming (or flowcharting) techniques. In the process, assign numeric labels to business functions.

7. Organize survey responses, interview notes, and other collected documentation by business function. This may be done manually, or using an automated tool, or by both methods. Prepare a file coversheet for each business function file.

8. Prepare an Aggregate Outage Impact Costs worksheet using data collected on the Business Function Survey. Graph preliminary estimates of dollar costs for business function interruptions of 24-, 48-, and 72-hour duration.

9. Perform a preliminary function criticality assessment based on survey responses. Revise the assessment based upon interdependencies between business functions, expressed as inputs and outputs.

10. Prepare a statistical summary of critical, vital, and nonvital functions.

11. Obtain documentation on automation resources used to support business functions. Identify technical data collectors that may be installed in company systems and networks and learn about the information these collectors can provide.

12. Work to relate automation resources to specific business functions to identify opportunities for consolidation.

13. Consult records management experts and user questionnaires to identify records storage and retention requirements.

Exposure and Risk Analysis

Risk analysis is the most misunderstood activity in disaster recovery planning. The methodology of risk analysis is distantly related to the type of analysis conducted by financial investors and insurance companies. However, as in these related fields, mathematical precision is not the goal of risk analysis in contingency planning. Instead, mathematics—or more to the point, statistics and probabilities—serves as a convenient shorthand for representing intangible values in conjunction with known quantities.

Having stated this, it is important to give credit to the salient points in the arguments of risk analysis' detractors. It is true that quantitative methods are often used to represent the probability of disaster, the exposure of the company to losses from disasters, and the benefit to the company of investments in disaster prevention and recovery. Figure 5.1 provides an overview of some of the formula recommended by risk analysts in the security and disaster recovery planning fields. These cost-justifications or return-on-investment calculations are, by their nature, misleading.

To establish a probability statistic, one requires an extensive empirical data set that identifies disaster incidents by type, duration, and impact (dollar cost). This data set does not exist, even within the business insurance industry. Hence, probability factors are statistical contrivances.

Additionally, even when extensive data sets are available, as in the case of 100 years of records on hurricanes, the best that may be said is that a disaster may occur at some point in the coming year and within some geographical range. Of what use is this data? Knowing that there is a high probability that three hurricanes will strike the Eastern Seaboard of the United States in 1999 is without value in the context of cost-justifying a disaster recovery plan for a company in Charlotte, North Carolina.

RISK ANALYSIS WORKSHEET
QUANTITATIVE FORMULA

ESTIMATING LOSS POTENTIALS

SIMPLIFIED LOSS EXPECTANCY (Quantify loss based on "rate of threat" and average single incident cost)

Tr = Rate of threat occurance
Sl = Single incident cost

$$LOSS = (Tr) * (Si)$$

ESTIMATE OF POTENTIAL LOSS (Quantify loss based on probability of occurrance and tangible/intangible
costs)

Po = Probability of Occurance (1% or .01)
Dc = Degree of Confidence
Co = Direct cost of outage
Cr = Cost for recovery
Ci = Cost of intangibles
EPL = Estimated Potential Loss

To estimate loss potentials:

$$EPL = Po * Dc * Co + Cr + Ci$$

ANNUAL LOSS EXPECTANCY (Quantify annual loss exposure based on probability and frequency of outages with a known or estimated cost)

Le = Loss Estimate
Po = Probability of Occurence
Fo = Frequency

To calculate ALE:

$$ALE = (Le) * (Po) * (Fo)$$

Figure 5.1 Quantitative formula.

ANNUAL LOSS ESTIMATE BASED ON EFFECTIVENESS OF PREVENTIVE MEASURES

Le = Estimated loss due to an outage
Ep = Effectiveness of preventive measures (expressed as a percent)
Cp = Cost of preventive measures

Calculate ALE:

$$ALE = (Le) - ((Le) * (Ep)) + Ce$$

DEMONSTRATING BENEFITS

SIMPLE BENEFIT ANALYSIS

Ln = Expected Loss due to Interruption without Disaster Recovery Plan
Lp = Expected Loss due to interruption with Disaster Recovery Plan

To calculate purported benefit of disaster recovery capability:

$$BENEFIT = (Ln) - (Lp)$$

BENEFIT AS REDUCTION IN ANNUAL LOSS EXPECTANCY

Le = Estimated loss due to an outage
Ep = Effectiveness of preventive measures (expressed as a percent)
Cp = Cost of preventive measures
ACS = Annual Cost Savings

Calculate ACS

$$ACS = (Le) - ((Le) - ((Le) * (Ep)) + Ce)$$

Figure 5.1 Quantitative formula. (Continued)

Noting the absence of reliable probability statistics, some risk analysts prefer not to mention probabilities at all. They instead refer to a "rate of threat" or a "frequency of interruption." These are merely probability factors masquerading as something else. That the operating logs of the corporate mainframe identify 25 incidents of downtime due to equipment problems in the last year is not germane to disaster recovery risk analysis, at least not statistically. Unless these incidents resulted in disastrous interruptions of automation services (in which case the best disaster recovery plan would be an up-to-date resume), the best that could be said from this data set is that approximately the same number of inconvenient interruptions would probably occur in the next year. There is no evidence, based on incident rate or frequency, that any of the interruptions would have disastrous outcomes.

The statistical fallacies of quantitative risk analysis have led many to dismiss risk analysis altogether as a transparent effort to baffle gullible management with scientific-sounding numbers in order to secure funding for security and contingency planning. Even generous critics regard quantitative risk analysis as an ill-fated attempt by technically minded disaster recovery planners to sell the need for planning to management using terms and language they believe to be familiar to business managers.

However, developing statistical models is not the only—or even among the important—activities of risk analysis. As indicated in Figure 5.2, risk analysis incorporates the data collection activities discussed in Chapter 4 and continues data collection efforts to better understand the threats to company operations, the potential costs to the company of interruptions, and the strategies employed by other companies to protect themselves against disasters.

In the process, of course, there is also an effort to build a case for developing a disaster avoidance and recovery capability. This case is based upon loss exposures, expressed as potential dollar losses resulting from a disaster, as well as legal mandates and plain old common sense. This case is presented to senior management at the conclusion of risk analysis, together with a budget proposal for the development phase of the disaster recovery planning project. If management is disposed to allocate the necessary resources for further plan development, the next set of activities will commence.

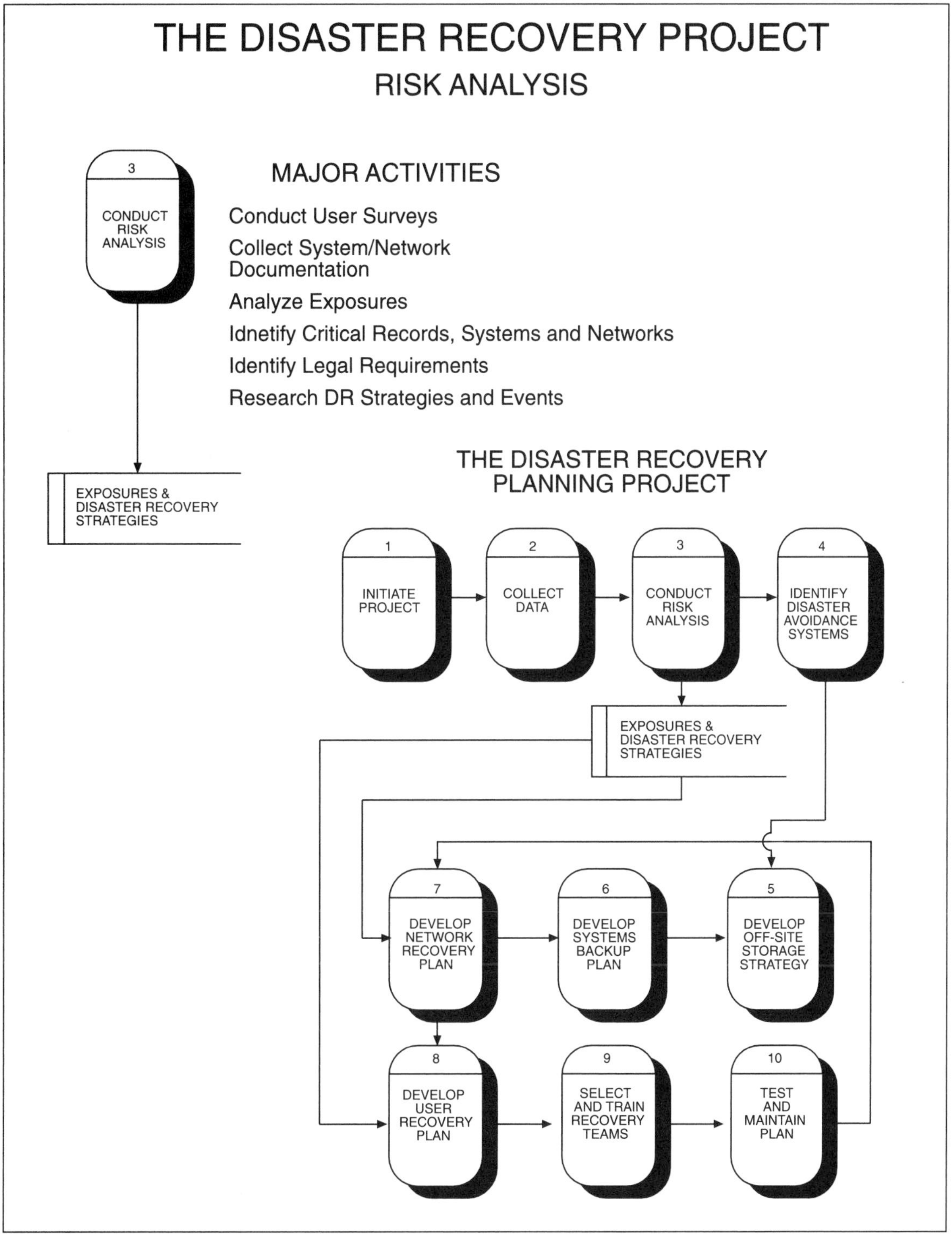

THE DISASTER RECOVERY PROJECT
RISK ANALYSIS

3

CONDUCT
RISK
ANALYSIS

MAJOR ACTIVITIES

Conduct User Surveys

Collect System/Network
Documentation

Analyze Exposures

Idnetify Critical Records, Systems and Networks

Identify Legal Requirements

Research DR Strategies and Events

EXPOSURES &
DISASTER RECOVERY
STRATEGIES

THE DISASTER RECOVERY
PLANNING PROJECT

1
INITIATE
PROJECT

2
COLLECT
DATA

3
CONDUCT
RISK
ANALYSIS

4
IDENTIFY
DISASTER
AVOIDANCE
SYSTEMS

EXPOSURES &
DISASTER RECOVERY
STRATEGIES

7
DEVELOP
NETWORK
RECOVERY
PLAN

6
DEVELOP
SYSTEMS
BACKUP
PLAN

5
DEVELOP
OFF-SITE
STORAGE
STRATEGY

8
DEVELOP
USER
RECOVERY
PLAN

9
SELECT
AND TRAIN
RECOVERY
TEAMS

10
TEST
AND
MAINTAIN
PLAN

Figure 5.2 Risk analysis phase.

ASSESS THREATS

Among the first tasks of risk analysis is threat assessment. This task seeks to acquire data on the potential causes of disastrous interruptions so that they may be considered within the context of the company's physical location, support infrastructure, and prevention capability.

Information on threats is available from numerous sources, some of which were cited in Chapter 3. In addition to threat data, valuable insights into threat potentials may be gleaned from disaster case histories found in the computer trade press and in magazines and journals focusing on computer security and disaster recovery planning.

Different Classification Schemes

For organizational purposes, it may be beneficial to classify threats by category or type. Many classification schemes have been advanced by planning professionals and others over time that may have caused some of the confusion that novice planners typically confront.

- **Natural versus Man-made Disasters**—One classification scheme that has probably evolved from the law of contracts divides the disasters into acts of man and acts of God or nature. This distinction is fuzzy at best. If flooding occurs in an office building due to heavy rains and a collapsed roof, is the cause of the ensuing disaster natural (the rain) or man-made (the faulty construction of the roof)?
- **Local versus Regional**—This scheme endeavors to classify disasters by their geography. If a company building burns to the ground, this is a local disaster; if this fire is part of a conflagration that consumes many buildings (following an earthquake or a volcano eruption or a riot), the disaster is considered regional. The problem with this division is that it begs the point: If any interruption of business functions is a disaster, what is the difference if fire has consumed one building or three? Every disaster is "local."
- **Cause of Disaster**—Derived from the sciences (and possibly from philosophy), this scheme seeks to classify disasters by their causes. Like the man-made versus natural disaster scheme, the causal approach is flawed. For example, if a fire guts a data center, the causal scheme would require more information about the origins of the fire for classification purposes. Was the fire the result of arson, equipment malfunction, human spontaneous combustion, or some other cause? For the threat to be classified, its cause must be identified.

Other schemes, including classification by severity, security versus contingency, and so forth, have been advanced from time to time by writers in the contingency planning field that help to muddy the minds of novice planners. The result of this confusion may be an inordinate amount of time spent in developing disaster scenarios that must be addressed by the disaster recovery plan. Strange as this may seem, many first-time planners (including this author) fall into the trap of dedicating months of work to creating loss scenarios for every conceivable type of disaster. The result is a massive written product that consumes the bulk of the planning budget before any recovery strategies are ever identified.

To avoid this trap, a simple, pragmatic method may be of use. Classify disasters by their effect.

Classification by Effect

The effects of any disaster fall into two distinct subsets. Either all business functions are interrupted or some business functions are interrupted. The duration of the interruption—how long the company will be deprived of the services or resources it requires to do business—is what qualifies the event as a disaster. The cause of the disaster is only important from the standpoint of certain insurance reimbursements and disaster prevention program development.

Two scenarios, therefore, are all that need to be considered from the standpoint of disaster recovery planning: total loss and partial loss. A *total loss* scenario involves any interruption that deprives all (or the majority) of critical and vital business functions of the resources they require. Hence, a failure at the telephone company serving office that interrupts incoming and outgoing communications traffic may be regarded as a total loss event despite the fact that all company resources and facilities are intact. The same may be said of a building fire that renders company quarters uninhabitable.

A *partial loss* scenario envisions the interruption of some critical business functions, while others are unencumbered. A roof collapse in an area of the company building where an important minicomputer is located may interrupt the operations of work units that depend on this automation resource. The impact is disastrous in the sense that critical business functions are impaired; however, the response required to recover from this disaster would likely be much less extensive than the response required to recover from a total loss scenario disaster.

The effect of a partial versus total loss threat classification scheme on plan design is to encourage modularity. The targets for recovery, business functions, should be the focus of discrete modules of the plan. In this way, the entire plan or any of its parts can be activated, depending on the nature and extent of the interruption. This modular approach has other advantages besides flexibility. For one, plan testing is much more readily accomplished. Also, plan modules organized by functions may be more readily maintained as business functions themselves evolve and change.

Disaster Potentials

As previously stated, there is merit to considering the possible types of disasters that may confront the company. In particular, it is useful to know about the common causes of disaster in order to identify worthwhile detection, alarming, and suppression systems that can help prevent avoidable disasters. A basic list of disaster potentials would have to include the following:

- **Flooding**—The intrusion of water (or other liquids) into work areas and equipment rooms is the most common cause of disasters.

- **Fire**—Resulting from a number of causes, fire damage to businesses totals in the billions of dollars annually. The life-threatening aspect of this disaster potential places it first on the threat lists of many disaster recovery planners.
- **Wind**—A natural disaster potential, high winds need not cause outages directly, but may create conditions for other disasters by damaging facilities, downing power or telephone cables, etc.
- **Infrastructure Outages**—Power drops or surges, loss of water or sewer services, air conditioning plant failures, and telephone line cuts are all infrastructure outages that can render company facilities uninhabitable or impair the operation of vital systems and networks.
- **Hardware and Software Failures**—System and network hardware is delivered with a set life expectancy, expressed as mean time to failure (MTTF) and mean time to repair (MTTR). Regular preventive maintenance may prolong the life of equipment, but wear, the effects of environmental contamination, and operator errors will eventually lead to the failure of even the most sturdy component. Software failures may result from sources too numerous to count.
- **Sabotage and Accidental Destruction**—Deliberate or unintentional damage may be caused to facilities or resources by company personnel, competitors, or innocent bystanders.

Environmental Considerations

Some disaster potentials need to be given greater weight in planning as a result of environmental considerations. For example, certain geographic areas are more susceptible to hurricane or earthquake damage than are others. The Federal Emergency Management Agency has prepared maps showing natural disaster risk areas throughout North America. They also maintain a special office for natural and technological hazards that can be of use as an information resource in identifying geographical disaster potentials.

In addition to natural disaster potentials, the proximity of a company facility to other firms that engage in hazardous manufacturing or processing activities or hazardous materials storage—such as nuclear power generation, paint and chemical processing, and so forth—may also need to be given special consideration by disaster recovery planners. These threat potentials can often be identified by speaking with local civil emergency management planners or fire protection agencies. Both organizations typically maintain information on potential local hazards for use in their planning efforts.

Civil emergency managers also need to be consulted regarding the provisions of emergency evacuation plans. Plans are typically developed and maintained by these agencies to cope with events such as hurricanes, earthquakes, and civil disorder. Planners will need to know under what conditions their facilities will be evacuated and what procedures exist for reentering facilities following the cessation of emergency conditions. These provisions will need to be accommodated in business recovery plans.

DEFINE EXPOSURE

A second objective of risk analysis consists of identifying the loss potential or exposure of the company in the event of disaster. The Business Recovery Survey discussed in Chapter 2 provides the views of work unit managers regarding their potential dollar costs in the event of an interruption. This data needs to be supplemented with estimated costs for fixed asset replacement or reconstruction, valuations of non-quantifiable costs, and other costs items such as legal liability, insurance deductibles, and late fees and penalties on financial arrangements.

Fixed (Replacement) Costs

Work unit managers are in a position to estimate the replacement costs for furnishings and office equipment used by their department. However, they probably (and this should be verified) have not included the replacement costs of centralized mainframe processors, PBX hardware, and other automation resources that are shared between several functions.

For the purpose of disaster recovery planning, dollar losses from these items may be expressed either as depreciated value or as the cost of replacement not covered by insurance. Planners may need to work with DP management or with the company comptroller or financial officer to identify the value of fixed assets so this number may be added to the aggregate potential loss estimate.

Another item to be considered in some businesses is the value of inventory on hand. An estimate must be prepared of inventory value for inclusion in the loss potential estimate. Also, supplies, such as preprinted forms, electronic media, and so forth, should also be approximated and their value included on the estimate.

Nonfixed (Reconstruction) Costs

Certain types of exposures cannot be valued according to replacement costs. For example, how does one calculate the cost of a piece of computer application software that is lost in a fire? A replacement strategy might use the original cost to develop the software as a basis for estimating its worth. Another approach might be to adjust development dollars to account for current salary levels of personnel who would be required to redevelop the lost software.

Neither of these adequately represents the value of the loss of the software. What is to be done while a critical piece of software is being recoded and tested? Is there some way to add the value of lost productivity to the overall cost for replacing the software? Furthermore, if the software provides a critical or vital business function without which the company is out of business, how can this contribution be valued?

The fact is that it cannot be valued. Other provisions must be made to safeguard copies of the software and source code to prevent their loss in a disaster. Backup and off-site storage provide the keys to ensuring that this type of loss is never experienced. However, backup programs are components of the disaster recovery plan and may not be initiated until approval has been granted to proceed with the disaster recovery plan. Thus, the potential loss of this software must be quantified and added to the aggregate cost estimate. For this reason, either of the two methods described at the outset (original development cost or adjusted development cost) or some other justifiable number may be entered.

This is not to appear arbitrary, but in the case of reconstruction costs for items such as custom application software, custom hardware, records, and data, planners need to assume that absolutely nothing will be done to safeguard these assets in the absence of planning. A reasonable estimate of the reconstruction costs of these items, therefore, needs to be made and calculated into aggregate loss exposure estimates to help build the case for management's support of disaster recovery planning.

Where possible, estimates should be supported by case histories or vendor quotes. In the case of records reconstruction, for example, a dollar figure may be obtained from a vendor of salvage services based on the actual salvage costs that accrued to one of their customers. In the case of data reconstruction, a ground-floor estimate of reconstruction costs may be obtained from a vendor of data entry services. Provide a sample record of a database and an estimate of the total number of records that would need to be rekeyed (assuming source document availability). The vendor should be able to provide a cost and time estimate for the work.

In reality, reconstruction of nonfixed assets may be impossible or possible only at astronomical cost. If a worst-case loss estimate is recorded for these assets, chances are that management will see the costs of recovery as so enormous that they will elect to do nothing about disaster recovery planning but make their resumes current and hope for the best. Confronted by more comprehensible numbers, they may be inclined to support efforts to reduce exposures through disaster recovery planning.

Other Intangible Costs

The only true non-quantifiable costs are such intangibles as the loss of market share or the loss of client confidence. Attempts to develop realistic estimates of these exposures are doomed at their outset.

However, if maintaining client confidence is a high priority for the company (as in the case of a just-in-time manufacturer or a stock brokerage), there must be supporting text in the presentation of risk analysis findings to emphasize the adverse impact of a disaster in this area and the potentially positive impact of having a tested disaster recovery plan as a means for inspiring client confidence. As for an estimate of lost sales, use the aggregate outage impact assessment calculations made by work unit managers to make the case.

Other Costs

Other costs for inclusion in the calculation of overall company exposures include legal liability, insurance deductible payments (see the section on risk analysis), and late fees or penalties on corporate financial arrangements. Estimates for the latter cost items may be obtained from corporate financial units or internal audit.

Legal liability may include fines and penalties accrued to the failure to provide legally mandated disaster recovery plans in certain industries. Liability may also result from contract and warranty provisions between the company and its clients or customers. These may be derived from the contracts or warranties themselves by the company's legal or finance departments.

Additional legal exposures may result from Torts-based lawsuits, in which the remedies available to plaintiffs appear to be limited only by the "generosity" of the court. Unless there are clear precedents for this type of suit in the company's recent history (or in the recent experience of companies within the same industry segment), it is probably best to exclude this area of legal exposure from exposure calculations in order to keep cost estimates on the ground.

THE TIME FACTOR

As previously stated, time is a multiplier of loss exposure. The longer that a company is without critical and vital business functions, the greater the costs of the outage and the less likely it is that full recovery will ever be achieved.

The timing of a disaster—the time of the day, week, month, or year that an interruption occurs—is also a determinant of the losses accrued to the company. An interruption that occurs over the weekend or over a holiday may not have the same disastrous consequences as the same interruption occurring at month-end or year-end processing.

Recognizing these points, many disaster recovery planners seek to determine a maximum acceptable downtime estimate, beyond which the interruption of business functions will become truly disastrous. Still others seek to identify critical periods during which an interruption would be particularly disruptive and costly in order to calculate for management an exposure estimate based on a worst-case scenario.

Maximum Acceptable Downtime

Maximum acceptable downtime figures by industry have been advanced in research studies conducted by the University of Minnesota and University of Texas at Arlington. According to the Minnesota study, financial companies have the lowest tolerance to interruption, a scant 48 hours, before the loss of business functions will begin to become nonrecoverable. The distribution industry can sustain its operations for 72 hours, while manufacturing concerns, insurance companies, and other industries may survive for up to six days following an interruption. This data represents an average based on estimates made by survey respondents.

The Business Function Survey administered at the outset of data collection may provide better information for the planner's use in calculating the maximum allowable downtime for the company. Questionnaire data, summarized on Preliminary Criticality Assessment Worksheets, may be quantified to show what portion of the aggregate cost estimate will accrue to the company in the first 24 hours, what portion will accrue in the next 24 hours, and then in the next 24 hours.

For the purpose of this book, 72 hours is considered the target timeframe for recovery. Some businesses may be able to withstand a more protracted interruption, while others (including many financial services companies according to the University of Minnesota) may realize the full financial impact of the disaster within a lesser period of time. The timeframe for recovery that is presented to management should match the requirements of the reader's company and not the 72-hour timeframe that is the current industry rule-of-thumb.

Timing of a Disaster

Case studies repeatedly demonstrate that the timing of a disaster can either speed up the accrual of disaster costs or slow them down. There is no reliable way to assign a value for the multiplier effect of disaster timing, however. This book, therefore, recommends that planners estimate interruption expenses without reference to the timing of the disaster and reserve discussions of the potential impact of disaster timing for the footnotes of the disaster recovery plan overview.

RISK ANALYSIS

Having identified disaster potentials and estimated company exposures, the disaster recovery planner may proceed to the third and final task of risk analysis: performing the risk analysis itself. As mentioned at the outset, this activity is frequently identified with the use of questionable quantitative formula that are deliberately or unintentionally weighted to favor the case for an investment in disaster recovery capabilities. In essence, quantitative formula are designed to demonstrate the probability of a disaster so that management is made to consider not only the question of "what" but also of "when."

Risk Management

Risk management is the overarching body of principles and techniques from which disaster recovery risk analysis is derived. Risk management rests on the calculation of probabilities and cost-efficiencies to identify how much is worth spending to safeguard investments or assets of a particular value.

Risk management is most effective where probability statistics are *hard* numbers. In the insurance industry, probabilities of death and illness have a hardness about them based upon an extremely large empirical data set. Millions of people become ill or die every year. This data is trapped and converted into actuarial tables that present, with a reasonable degree of accuracy (e.g., probability), that a 30-year-old, married male who smokes will live 53 years to the age of 83. Based on this data, a life insurance policy may be sold to the customer with a fair degree of certainty that his heirs will not be cashing in on the policy for many years.

Within the context of disaster recovery planning (and, indeed, in the business insurance industry), such actuarial data is not available. One reason is that business "life expectancy" is too short for reliable data to accumulate on the frequency and impact of interruptions. Currently, small business ventures have an average life span of about six years. Even with older firms, data is difficult to collect. Mergers and acquisitions muddy the data sample and, until recently, disasters were often a closely held secret.

In the absence of actuarial data, it is impossible to predict the probability of disaster. Industry rule-of-thumb places the chance of total loss-type disaster each year at 1 in 100 or .01. Even this statistic seems somewhat arbitrary.

How does the probability of a disaster figure into the rationale for undertaking such planning? Risk management principles provide the answer. To determine how much should be spent to safeguard an investment or asset of a known value, multiply the value of the asset by the probability of loss. The resulting number provides a ceiling for expenditures strategies for reducing risk.

In other words, if the aggregate loss estimate from a disaster is $1,000,000, and the probability of a disaster occurring this year that would result in the loss is 1 in 100 or .01, then the loss exposure this year is $1,000,000 X .01 or $10,000. Therefore, $10,000 would be regarded as the ceiling on expenditures to protect the asset.

Permutations on this basic formula (see Figure 5.1) may be used to cost-justify the acquisition of disaster prevention systems that will reduce annual disaster probabilities or to otherwise demonstrate the benefits of disaster recovery planning as a return on investment. Some planners argue that this sort of statistical justification is necessary to convince management to invest in a capability that produces no revenue and, in the best of circumstances, measures its success in non-events. Others contend that management will see through these statistical contrivances immediately, and that planners who employ these methods will damage their own credibility.

In addition to the questionable calculus for disaster cost-justification just cited, two other models for evaluating and presenting exposure data and for cost-justifying measures to reduce exposure are in widespread use at this time and thus merit attention here. One model is based upon a ranking of disaster threats based on an intuitive evaluation of their likelihood of occurrence by the members of the disaster recovery planning team.

A list of disaster potentials, such as the one in Figure 5.3, is given to each team member in what is called a *Delphi Session*. (This listing provides a framework for discussion, but is not strictly required in the Delphi approach.) Moving from one threat to another, the team discusses the relative likelihood of a disaster of the specified type occurring within the next year. A consensus is reached to assign a ranking—from one to five—to each disaster potential.

Based on the Delphi ranking, threat potentials may be arranged on a continuum, such as the one shown in Figure 5.4, indicating the frequency of the threat and its associated loss potential. This continuum is a representation of an intuitive fact of disaster recovery planning: Risks with high dollar loss potentials are generally the least likely to occur. Power outages may shut down operations on a monthly basis during the lightning storm season. However, the duration of the outages may not cause them to have disastrous consequences or to pose high dollar costs to the company. On the other hand, a direct strike by a hurricane that demolishes company facilities is likely to be regarded low frequency event with a high dollar cost potential.

THREAT RANKING WORKSHEET

INSTRUCTIONS: Evaluate this list of threats and assign each a ranking based on your estimate of likelihood of occurrence within a one-year timeframe. 1 is not likely, 5 is quite likely.

DISASTER POTENTIAL	RANKING: 1	2	3	4	5
WORK AREA FIRE					
DATA CENTER FIRE					
POWER INTERRUPTION					
WEATHER EVENT (HURRICANE, FLOODING, ETC.)					
INDUSTRIAL SABOTAGE					
CIVIL EMERGENCY (RIOTS, STRIKES, ETC.)					
HARDWARE FAILURE (TELECOMMUNICATIONS OR EDP)					
SOFTWARE FAILURE					
FACILITY BREECH (ROOF LEAK, LOSS OF A.C., ETC.)					
WATER STOPPAGE					
LOSS OF TELECOMMUNICATIONS GATEWAY ACCESS					
COMPUTER VANDALISM/VIRUS					
BOMB THREATS					
PLANE CRASH					
HEALTH EMERGENCY (NUCLEAR, CHEMICAL, ETC.)					

Figure 5.3 Threat ranking worksheet.

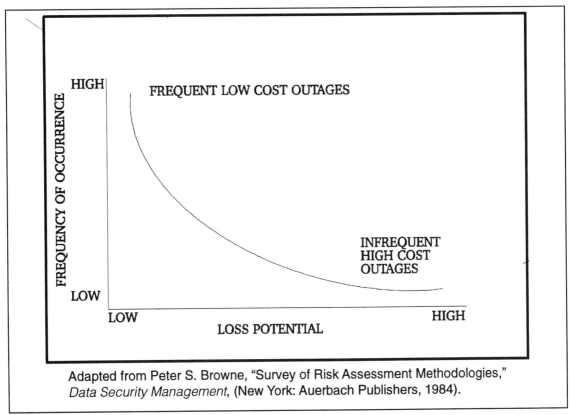

Adapted from Peter S. Browne, "Survey of Risk Assessment Methodologies," *Data Security Management*, (New York: Auerbach Publishers, 1984).

Figure 5.4 Ranking risks by frequency/loss potential.

The point of this evaluation is not to establish specific scenarios to treat in the plan, but rather to arrive at, by intuitive means, a qualitative case for supporting the planning effort. This effort, however, is not without its statistical contrivances.

Once threats have been assigned a frequency value, the aggregate costs associated with an outage (as previously calculated) are multiplied by the frequency statistic to identify an annualized loss exposure. This result is, in turn, used to identify a "break even" point for expenditures on disaster recovery planning, or in the words of one proponent, Peter S. Browne, to provide management with some means "to make a rational decision regarding controls."

Figure 5.5 shows the application of this approach. For a comparatively low price, a disaster preparedness capability may be obtained to offset those risks with the greatest likelihood of annual occurrence. Spending money to prepare for low frequency risks beyond this point may be viewed as excessive. Note that, in Figure 5.5, the curve for "expected loss" represents expected losses from risks with a high frequency of occurrence. A less likely risk, were it to materialize, may carry an extremely high cost. However, since the high cost risk is less likely to occur within the year, the annual loss expectancy from this risk is expressed as a low expected cost. The expense for protecting against a low risk threat is generally quite high.

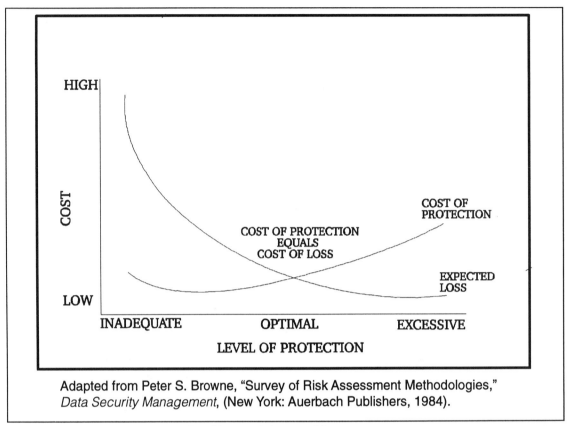

Adapted from Peter S. Browne, "Survey of Risk Assessment Methodologies," *Data Security Management*, (New York: Auerbach Publishers, 1984).

Figure 5.5 Disaster recovery spending: How much is enough?

Objections to the validity of the quantitative method used in the previous assessment and analysis aside, the end result is a fairly cogent model for rationalizing disaster recovery planning and for convincing management of the disaster recovery planner's attention to the bottom line. Currently, a second simplistic model is also popular for communicating the need for disaster recovery spending to management.

One disaster recovery services vendor claims that he invented the "window of exposure" model on the back of a bar napkin when seeking to cost-justify his disaster recovery product to the chief financial officer of a prospective client company. As shown in Figure 5.6, the "window of exposure" refers to the time and money that a company can expect to lose in the event of an outage. Here, there is no concern with the probability of a disaster occurrence within a given year, merely an acknowledgment of the fact that disasters happen. What is of concern is the raw, aggregate exposure of the company to loss in the event of a prolonged interruption of normal business functions. This is the exposure window that the company must quantify for itself.

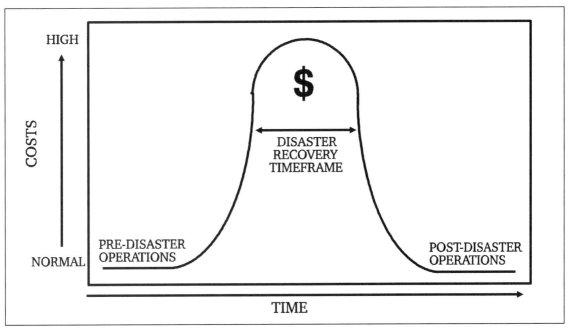

HIGH

COSTS

$

DISASTER
RECOVERY
TIMEFRAME

PRE-DISASTER
OPERATIONS

POST-DISASTER
OPERATIONS

NORMAL

TIME

Figure 5.6 Window of exposure.

Once the exposure window has been quantified, according to the adherents of this model, rationalizing expenditures for disaster recovery planning is simple. As shown in Figure 5.7, plan costs (the costs for purchases of disaster recovery products and services) represent an additional cost to be absorbed during periods of normal operations. However, these expenditures should reduce the duration and the cost of the exposure window by a proportional amount. Placed in arithmetic terms:

$$(X-Y) <= Z$$

That is, the difference in the dollar loss exposure before the acquisition of disaster recovery capabilities (X) and after the acquisition of these capabilities (Y) is quite simply the maximum amount that may be justified by planners in their requests for allocation of corporate monies for disaster recovery.

This model would likely be condemned by quantitative risk analysts for its lack of appreciation for the mitigating effect of probabilities on loss exposure. However, the model does have one or two distinct charms. First, it is not without some basis in fact. As illustrated in Figure 5.8, various disaster recovery strategies do yield different recovery timeframes and, thus, do represent different associated exposures. The model, if nothing else, clearly demonstrates to management the key benefit of planning: exposure reduction.

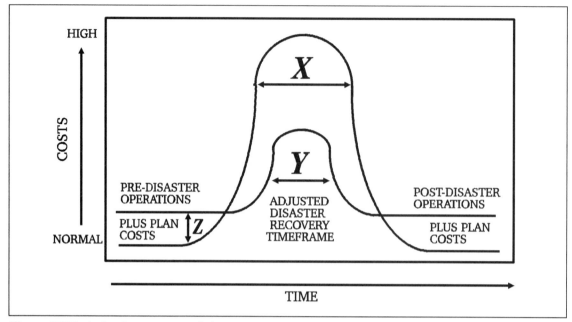

Figure 5.7 Reducing the exposure window.

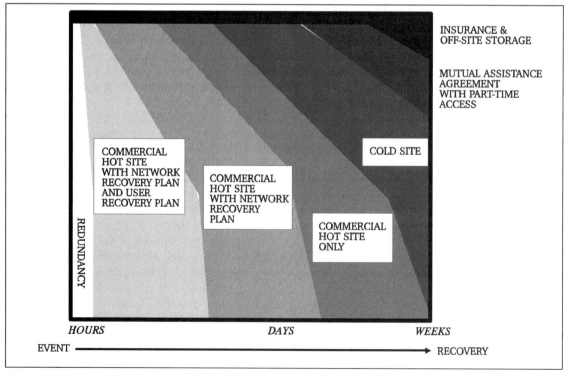

Figure 5.8 Recovery timeframes.

A Pragmatic Approach

Whether quantitative methods are used by the disaster recovery planner is a matter of choice. At best, these models provide interesting tools for discussing what might otherwise be perceived as a tenuous compilation of disjointed facts and opinions. At worst, models will be presented as factual, exposing their presenter to scorn and ridicule should a savvy manager ask to see the basis of the probability figures being used.

What must be acknowledged by the planner is that the presented data is unconventional, yet intuitively correct. The planner must also recognize that a worst-case, total-loss scenario is extremely unlikely to materialize. Quantitative formula may be advanced, with carefully caveated assumptions, but only as descriptive aids, not as predictive oracles. Ultimately, it will be in the hands of management to decide whether they will allocate X, Y, or zero dollars to safeguarding their business functions.

Planners who are honest about the strengths and the shortcomings of risk analysis have the best chance of succeeding in obtaining management support. The fact is, based on case studies, that companies that plan for the possibility of a disaster tend to recover from disasters. Those that do not plan, do not recover.

INSURANCE AS A DISASTER IMPACT MITIGATOR

Many business managers operate under the illusion that disaster recovery planning is unnecessary if the company maintains the right mix of business insurance coverages. They often cite Director and Officer (D & O) Insurance as the stopgap that prevents a corporate emergency from becoming a personal one. In other words, they are confident that D&O Insurance will protect them from any deep pocket lawsuits brought by stockholders or customers who argue that the company's management did not do enough to prepare for and minimize the impact of disasters.

The fact is that insurance can play an important role in the financing of a recovery effort and it does mitigate some of the costs accrued to disaster if properly purchased. Therefore it is important to know what coverages are already purchased by the company before presenting the findings of the risk analysis.

Types of Coverages

There are several types of insurance coverages that should be considered by any company that relies on automated systems and networks to engage in business. These include business interruption insurance, EDP insurance, special coverages, and self insurance.

Business Interruption Insurance

Business interruption insurance is a popular form of coverage designed to reimburse (or prepay) specific losses predicted from an unplanned interruption of normal business operations for a set period of time. Covered losses may include continuing expenses for doing business from an alternate location, key employee salaries, and other expense items known in advance and specifically included in the policy.

It is important to note that knowing specific cost items in advance is an asset when purchasing this type of coverage. Also, firms that provide business interruption insurance state that it is much easier to find the right mix of cost and coverage for a client company that has thought about and planned for the actions to be taken in the wake of an unplanned interruption.

EDP Insurance

As its name implies, EDP insurance is a specialty policy to cover electronic data processing equipment. These policies, which can also be purchased to cover telecommunications and even microcomputer systems, are available in three classes. Coverage may be purchased to cover equipment replacement or repair expenses. A second class of coverage may be purchased to cover the cost of re-entering data into damaged databases using extant records and a crew of data entry clerks. A third class can be used to cover extra data processing expenses, such as the declaration fee that the company will have to pay to activate its contract with a data processing hot site.

In the case of EDP insurance, as with business interruption insurance, knowing what coverages to purchase becomes easier if disaster recovery planning has been undertaken. Again, insurance companies prefer to work with firms that have calculated risks and exposures as part of a disaster recovery planning project. This is particularly the case in the negotiation of extra expense coverages—special coverages for preplanned expense items related to EDP restoration or continuity. Extra expense coverage may be prearranged for such items as 24-hour, on-site, technical support from a vendor in the wake of an equipment outage or for overnight mailing of critical equipment or data to an alternate operations facility.

Special Coverage

Similar to extra expense coverages in EDP insurance policies, businesses may also purchase special coverage insurance for disaster potentials that are beyond the scope of a basic standard business interruption insurance policy. Special coverage insurance may be purchased to broaden covered disaster potentials, such as fire and theft, so that policy coverages include losses due to roof collapse under the weight of snow, ice damage, broken windows, and so forth.

Self Insurance

Another popular insurance strategy is the laissez faire approach of self insurance. As the name implies, some companies elect to cover their own losses in the wake of an unplanned interruption following an analysis of risks and mitigative strategies. Rarely is this option possible for publicly held firms or companies with debts such as business loans or mortgages. But, a surprising number of private firms still use this strategy to keep business costs to a minimum.

Advantages of Disaster Recovery Plan in Insurance Planning

As previously stated, companies stand to benefit from disaster recovery planning in advance of business insurance purchases. Firms that have planned for business interruption, and that have articulated strategies and procedures for coping with disaster, are generally better informed about their own insurance requirements. Insurance can be purchased in a more targeted way to facilitate the economic requirements of the recovery.

Representatives of insurance companies are quick to point out that the prices for their policies are not necessarily reduced if a customer has a disaster recovery plan. However, coverage can be purchased more rationally in the presence of preplanning, thereby reducing waste and lowering costs overall.

Role of Insurance in Disaster Recovery

Insurance can provide an important part of the financing required by a company that is responding to an emergency. Figure 5.9 provides a useful format for identifying insurance policies maintained by the company and their coverages for purposes of planning for recovery financing.

However, many observers point out that insurance should not be the only source of emergency funds. While insurance companies may respond quickly to a disastrous interruption resulting from a telecommunications switch damaged by an electrical power surge, they are also notoriously slow in responding to other disaster scenarios.

For example, many insurers will not pay on policies involving fires until the fire department has conducted an investigation and released the "crime scene." Also, as recent natural disasters have demonstrated, insurers can slow to a halt when evaluating large numbers of claims received from businesses in a specific geographic vicinity.

Insurers are quick to point out, however, that companies can integrate procedures into their disaster recovery plans that may help to expedite claims processing. These procedures range from contacting the right persons in the insurance company bureaucracy and utilizing the correct forms for claims submission to prearranging for witnesses to be deposed at the scene or photographers to make a full record of damages. The common wisdom is that claims processing timeframes will be reduced by any procedures the company implements in advance to facilitate the resolution process.

INSURANCE POLICY WORKSHEET

INSTRUCTIONS: Complete this form to identify business insurance policies that may be utilized in the wake of an unplanned business interruption. Enter the policy number, vendor, type(s) of coverage provided, and contact information for activating policies should the need arise.

Policy Number	Vendor	Type of Coverage	Special Provisions	Contact Information

SUMMARIZE EXTRA EXPENSE OR SPECIAL COVERAGES PROVIDED:

_____

_____

_____

_____

_____

Figure 5.9 Insurance policy worksheet.

PRESENTING FINDINGS

The presentation of the outcome of risk analysis proceeds along the same lines as the initial presentation detailed in Chapter 3. A preliminary findings report may be issued and a formal meeting date set. The agenda for the presentation will include:

- **Current Exposures and Projected Loss Impacts of a Disaster**—This topic will cover the findings of the data collection and risk analysis efforts and will describe the methods used to collect and evaluate data.
- **Objectives of the Development Phase**—This topic examines the specific business functions (critical or vital) that will be targeted for full recovery, the timeframe for recovery, and the approach that will be taken to develop strategies for recovery.
- **Parameters on Strategy Development**—This topic provides the opportunity to express the project team's attention to the ancillary goal of cost-efficiency in all recovery strategies. Quantitative models may be introduced, with appropriate caveats, to rationalize costs, and other benefits of disaster recovery-related expenditures (i.e., reduction in insurance costs, marketing value of preparedness, etc.) may be re-emphasized.

The presentation just outlined may be conducted immediately following risk analysis or delayed until after disaster prevention strategies have been identified (see Chapter 4). The advantage of waiting until disaster prevention systems have been identified and priced is that the planner may receive approval to acquire and install prevention systems whether the further development of the disaster recovery plan is approved. In companies that lack any safeguards, this represents a major contribution.

If the presentation is made prior to the identification of disaster prevention strategies, no budgetary requests—save for a prolongation of current planning budgets—are needed. All products and services sought for use in disaster avoidance or recovery capabilities will be submitted to vendors for bid and a budgetary request for their implementation proposed at a later time.

SUMMARY

Risk analysis is a common area of misunderstanding and difficulty for novice planners. Many become bogged down in endless scenario-building and analysis. Others seek to build complex statistical models for event occurrence and ramifications.

It must be kept in mind that the quantification of risks and the purported reduction in corporate exposures due to disaster recovery expenditures are strictly theoretical constructs. Senior management will not be persuaded to support a disaster recovery planning budget based on imaginary returns on investment. Planners who present their risk analysis findings as conclusive demonstrations of statistical or cost benefits run the risk of being laughed out of the room.

Still, risk analysis has the merit of suggesting to senior management that disaster recovery planning is a serious business. Presentation of risk analysis results can demonstrate that the planner is sensitive to budgets and is trying to present a business case in language that managers understand. The planner does this in order to address a requirement that is intuitively understood by everyone.

Checklist for Chapter

1. List disaster potentials and assign a value to indicate their likelihood of occurrence.

2. Estimate the dollar cost of an unanticipated outage of 24, 48, and 72 hours.

3. Identify existing business insurance coverages. Identify opportunities for reducing costs and increasing coverages.

4. Calculate a cost-justifiable dollar amount to allocate to disaster recovery planning.

5. Present findings to management.

Facility Disaster Prevention Capabilities

RATIONALE AND OVERVIEW

Traditionally, disaster recovery planning aims at minimizing disaster loss potentials through the development of capabilities and procedures for handling unplanned interruptions of critical business functions. Such planning is often perceived as an essentially reactive process. The reactive nature of disaster recovery planning is implied in the term *recovery*.

However, the enterprise of disaster recovery planning is actually proactive in nature. Planning is undertaken prior to a disaster to establish and implement capabilities for coping with a disaster before one occurs. An important component of this proactive effort is planning for the acquisition and installation of systems for preventing avoidable disasters.

The fact is that many disasters do not result from a single cataclysmic event, but instead develop over time. Responding to a disaster potential before it becomes a disaster is the first objective of disaster recovery planning. The second objective is to minimize the costs associated with disaster potentials that cannot be eliminated.

To accomplish the first objective, planners must consider the likely causes of disaster, which is one goal of risk analysis. Next, they must research what means are available to eliminate or minimize the chances that a particular disaster potential will be realized. Having discovered appropriate risk reduction technologies, planners then need to rationalize the acquisition costs for these technologies and solicit management approval. Finally, they must work within the context of normal operational schedules to arrange for technologies to be acquired and installed. This chapter examines these steps in considerable detail, since they are among the most important in disaster recovery planning.

Why include the identification and acquisition of disaster prevention systems in a book for disaster recovery planning? The reason is simple. Many companies begin disaster recovery plans that are never completed and tested. Whether due to a loss of management support, belt-tightening and cost-cutting at the firm, or a host of other factors, more than a few companies report that their well-intentioned planning efforts were derailed at one point or another and never resulted in a finished plan.

It is the contention of this book that a disaster recovery capability is built in increments. For each smoke detector, fire suppression system, access control, or other preventive measure that is installed as a consequence of the disaster recovery planner's efforts, personnel and resources are better protected. This is not to say that the disaster recovery plan is unimportant, only that it is one part of a constellation of parts that together comprise a safer, more secure workplace. Therefore it is important to secure as much in the way of prevention systems and technologies as possible—if only as a hedge against the possibility that the formal disaster recovery project does not yield a workable plan.

Purpose of Installing Disaster Prevention Capabilities

Identifying disaster avoidance systems is a discrete node of the disaster recovery project model introduced at the outset of this book. As shown in Figure 6.1, this set of activities includes the analysis of exposure to *avoidable* disaster potentials and the identification, selection, and acquisition of suitable preventive technologies. The goal of these activities is to provide the company with a prevention capability that will reduce the risk of unplanned downtime and the exposure of employees to health and safety hazards.

Several disaster potentials are the targets for investigation and analysis at this stage of disaster recovery planning.

- Facilities, including user work areas, data centers, and switch rooms, will need to be assessed for their safe and secure design.
- The power distribution systems throughout the company, but especially in the data center, will need to be reviewed for their adequacy and resilience to known interdiction factors.
- Fire and flood detection, alarm, and suppression systems (or the need for these systems) will need to be identified.
- Existing physical access controls and electronic access controls must be assessed from two perspectives. Planners need to judge the effectiveness of these controls in reducing security risks. Also, planners need to know how security controls may be circumvented to accomplish recovery objectives in the wake of a disaster.

These potentials and corresponding criteria that may be used to assess the current level of preparedness at the company are summarized in Figure 6.2. Methods for assessing preparedness and for identifying areas for improvement are provided in the pages that follow.

THE DISASTER RECOVERY PROJECT
IDENTIFY DISASTER AVOIDANCE SYSTEMS

MAJOR ACTIVITIES

Identify Fire Prevention Requirements

Identify Flood Detection Requirements

Assess Environmental Contamination Level

Analyze Power Backup Requirements

Identify Physical Security Requirements

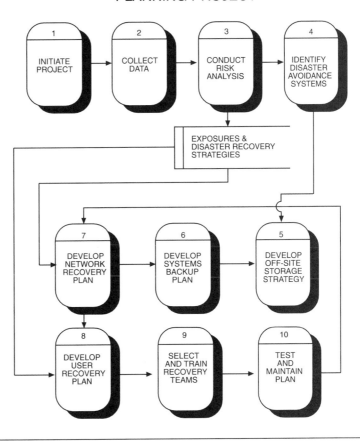

Figure 6.1 Disaster avoidance phase.

FACILITY CONSTRUCTION

Does facility design meet NFPA standards for
equipment rooms and work areas?

POWER DISTRIBUTION SYSTEMS

Are power distribution, control, and backup systems
adequate for load requirements?

FLOOD DETECTION & ALARM

Have potential flood sources been identified and
are detection and alarm systems installed?

FIRE DETECTION & SUPPRESSION

Are fire exits clearly marked? Are regular evacuation
drills conducted? What detection and suppression
systems are installed?

PHYSICAL ACCESS CONTROLS

What controls are in place to restrict physical
access to sensitive data, equipment, or work areas?
Can access controls be overridden in an emergency?

ELECTRONIC ACCESS CONTROLS

What controls exist to restrict access by
unauthorized persons to data, systems and networks?

Figure 6.2 Disaster prevention assessment criteria.

Other Advantages

The acquisition of capabilities for preventing fire, preventing power-related equipment failures, or restricting computer vandals from access to company systems and networks is clearly justified within the context of disaster recovery planning. However, unlike other disaster recovery planning expenditures, these capabilities can offer several side benefits to the company that may aid in cost-justifying their acquisition to senior management.

Two side benefits are readily apparent. First, insurance companies tend to give credit in the form of a reduced annual premium for nearly any preventive capability installed by the company. Secondly, some preventive systems may actually support and make more efficient the normal operational activities of the company.

Reduced Insurance Premiums

The installation of a fire prevention system or an uninterruptible power supply (UPS), as well as the existence of a documented disaster recovery plan, may reduce annual premium payments for business insurance. While actual insurance benefits are discussed in greater detail in Chapter 5, some generalizations may be offered here to justify this assertion.

In theory, insurance companies calculate client policy premiums on the basis of perceived risk. In the absence of a reliable statistical basis for evaluating risk, most insurance companies that offer policies to business firms—especially in the area of electronic data processing (EDP)—rely upon a combination of internal guidelines and on-site assessments by their agents as a means for rationalizing coverages and their costs. A high-risk venture may be unable to obtain insurance or it may be charged a stiff premium and saddled with a high deductible to enable the insurer to justify the insurer's exposure to policy payouts.

Generally, an EDP insurer looks for evidence that a prospective client has taken steps to reduce its own risk of EDP systems failure through disaster recovery planning and security planning. If insurance agents are persuaded that the risk of loss has been reduced by the development of a tested disaster recovery capability, they will recommend that insurance premiums and deductibles be reduced accordingly.

Disaster prevention systems, such as fire alarm and suppression systems, testify to a disaster prevention capability within the company and also to management's concerns for its own risk management. This encourages insurance agents regarding the acceptability of the risk.

Moreover, as will be demonstrated later, a disaster recovery plan is a necessary precondition for making intelligent insurance product purchases. Without a plan for coping with a business interruption, how can anyone purchase appropriate coverages for coping with an outage? Most companies that have business interruption coverages before disaster recovery planning is undertaken soon find that they have been spending entirely too much money in premiums for unnecessary coverages or that they have very little actual coverage at all.

Thus, the costs for acquiring disaster prevention systems, and the costs for disaster recovery planning generally, may be partially offset by reductions in business insurance premiums. How much or how little premiums will be reduced depends upon the judgment of the insurance agent.

Support for Daily Operations

The other potential side benefit of disaster prevention is its potential to make normal, day-to-day operations of the company more efficient. For example, Figure 6.3 provides an overview of some of the factors justifying the acquisition of an uninterruptible power supply (UPS). The UPS actually serves two purposes. Obviously, it reduces the likelihood of unplanned downtime due to equipment wear, data loss, program bugs, and other disaster potentials associated with line noise, transients and spikes, and power blackouts. It does so by providing sufficient battery power to meet the requirements of connected equipment until a blackout is past or until a generator can be brought on-line to carry the load.

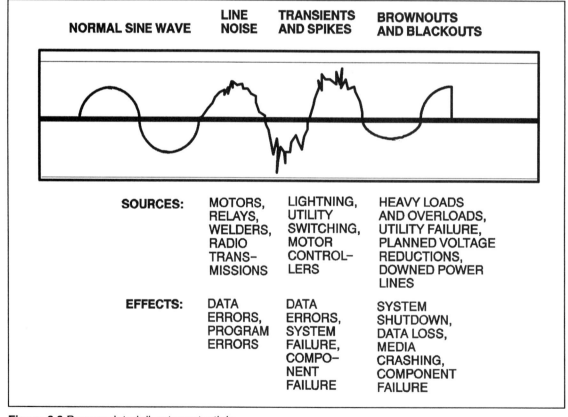

Figure 6.3 Power-related disaster potentials.

However, this same capability may make operations more efficient by surmounting minor outages and interruptions that may occur every time there is a lightning storm. More than one operator can tell the tale of a 20-hour batch job that reached the nineteenth hour of processing when a lightning strike caused a momentary loss of power. When the power was restored, the job needed to be restarted from the beginning. Clearly, a UPS would provide the means to sustain normal operations under these circumstances and would provide an improved efficiency in data center operations.

Another example of a practical side benefit of disaster prevention systems is demonstrated in a no-smoking policy, often implemented as part of a fire prevention program. A no-smoking policy will not only reduce the risk of accidental fire and minimize the levels of environmental contamination that can shorten the lives of some data processing equipment, but also have the added advantage of creating a more healthful workplace, reduce health benefit expenses, and—if statistics are correct—reduce the number of employee sick days annually.

Where possible, these practical benefits of disaster prevention systems should be included in the cost-justifications for their acquisition. Combined, insurance premium reductions and operational cost-savings may make a compelling case on their own for justifying the implementation of certain risk-reduction strategies and prevention technologies.

SCOPE

Reducing disaster loss potentials should be a factor in nearly every business project. From new facilities planning to system development, attention should be paid to disaster prevention and disaster recovery. Perhaps, as a better understanding of contingency planning requirements is disseminated through corporate organizations and expressed in the form of company policy, this basic principle will become part of how businesses conduct business. For now, however, it is likely that a disaster recovery planner will confront an environment that is ill-prepared to cope with the possibility of a disaster. It is in this imperfect environment that disaster prevention capabilities must be implemented.

Retroactive prevention entails certain costs. Facilities may need to be redesigned, procedures may need to be changed, and work flows may need to be altered—if only to install prevention systems. Thus, these costs, together with the acquisition costs for the systems, must be managed effectively.

This standard encompasses facility-specific prevention systems. In truth, however, a disaster prevention capability would require much more. Companies would be wise to articulate policies on security and disaster awareness, to formulate and enforce no-smoking areas, to require regular briefings by work area managers on the state of readiness of their respective work units, and so forth. These aspects of disaster prevention are beyond the scope of this book, however, because of the wide variance among organizations in terms of structure and political culture.

Suffice it to say that planners should attempt to frame their prevention strategies within a more global context by seeking policy edicts from management. However, they must use their own judgment about their organization and its management to determine how much can reasonably be sought and to avoid cultivating hostility or ill will.

General Facilities

The structure, design, and location of a facility are key considerations in the evaluation of risk potentials. Usually, disaster recovery planners do not have the luxury of selecting a facility site or a hand in designing it for disaster prevention. They must do their best to safeguard facilities that exist.

This is not to say that the planner's hands are completely tied. Construction codes, encompassing building materials, electrical wiring, fire prevention, and other factors, are enforced in nearly every U.S. state and municipality to govern how buildings will be built. An assessment should be made of facilities in this planning stage both to identify any code violations so they can be rectified and to identify opportunities for providing additional protection not specified in the codes.

Figure 6.4 is a worksheet that planning teams can use when inspecting work areas of the company for disaster prevention requirements. Ideally, planners performing inspections will be joined by a fire safety official or building inspector recruited on an informal basis. This way the planners can profit from the expertise of the officials without inviting sanctions or citations. If the official notes a code violation, this may be brought to management's attention so that action can be taken to resolve the violation before a formal inspection of the premises is made.

As noted on the form, inspections should seek general information about electrical power distribution and use, environmental factors, fire alarm and suppression capabilities, water (and sewer) damage potentials, and access controls. Information is collected through observation, employee interviews, and manager interviews for later use in rationalizing the investment in disaster prevention capabilities.

The Remarks section of the form should be used to note any special observations by experts who accompany the inspection team. For example, if a glass partition separates the raised floor area of a data center from a user work area, and a question is posed about the viability of chemical fire suppression systems in the raised floor area, these concerns and suggested remedies should be noted.

Another type of information that should be solicited from government representatives concerns man-made or natural hazards that may be present within the vicinity of the company facility. If chemical processing plants or nuclear facilities are located near the company site, planners will need to consider mandatory building evacuation as a disaster potential. For example, if Acme Chemical has a fire and begins releasing noxious fumes into the air, local health and safety officials may need to evacuate area businesses and homes for a protracted period of time until the fire can be put out and testing can be performed to ensure that any health risk has passed. If a business within the evacuated perimeter can expect to be denied access to its facilities for several days, it may need to activate a disaster recovery plan and temporarily relocate its operations to a backup data center and user recovery facility. This potential, which is a by-product of facility location, needs to be considered in the disaster recovery plan; local inspectors are usually in the best position to identify such threat potentials in the neighborhood.

DISASTER PREVENTION INSPECTION WORKSHEET

DATE OF INSPECTION: _____

AREA INSPECTED: _____

CONDUCTED BY: _____

IS THIS A REINSPECTION? YES ___ NO ___ (Check one) IF YES, INDICATE LAST INSPECTION DATE: _____

WERE MANAGERS GIVEN ADVANCE NOTICE OF THE INSPECTION? YES ___ NO ___ (Check one)

POWER

YES/NO

		Are overloaded outlets in evidence?
		Are loose extension cords in traffic paths?
		Is there any evidence (acrid smell, discoloration around outlets) to indicate recent overloads or short circuits?
		Are fused power strips in use?
		Do power cords and power strips bear UL stickers?
		Is any equipment protected by surge protection devices?
		Is any equipment protected by uninterruptible power supplies?
		Are wall circuits shared with load shifting equipment, such as refridgerators or air conditioners?

REMARKS:

ENVIRONMENT

YES/NO

		Is cigarette smoking permitted in the area?
		Was the area visited particularly warm or cold, or humid?
		Was positive pressure maintained in the area (did air blow out when you opened the door)?
		Were windows open?
		Did the room air smell musty or damp?

REMARKS:

Figure 6.4 Disaster prevention inspection worksheet.

FIRE ALARM/SUPPRESSION

YES/NO		
		Are ceiling sprinklers, standpipe and hose, or fire extinguishers in evidence?
		Are inspection labels on extinguishers current?
		Are fire exits clearly marked?
		Are smoke detectors installed? If so, do they respond to tests?
		Are hot plates, individual coffee machines, or heating elements in use?
		Are floor heaters in use?
		Are any fire awareness posters in evidence?
		Can an employee tell you the approximate date of the last fire drill conducted in the area?
		Are room walls fire rated?
		Are wastepaper bins or printer paper near electrical fixtures or equipment?
		Does any equipment feel unusually warm to the touch?

REMARKS:

If the subject area is a data center or telecommunications switch room, is HALON or some other non-water fire suppression system installed? YES ___ NO ___ If YES, indicate make, model, agent and date of last inspection or test:

WATER

YES/NO		
		Do restrooms abut the subject area?
		Is recent water damage evident on walls, floors, or ceilings?
		According to building plans, do water, sewer, or drain pipes pass through ceilings or walls of the subject area?
		Is water-cooled equipment in use?
		Do sprinklers or standpipes show condensation?

REMARKS:

If the subject area is a data center or telecommunications switch room, is some type of water detection system installed? YES ___ NO ___ If YES, indicate make, model and date of last inspection or test:

Figure 6.4 Disaster prevention inspection worksheet. (Continued)

YES/NO		ACCESS CONTROLS
		Does the work area inspected deal with sensitive or secure information?
		Are photo IDs required for access and are these IDs checked?
		Are terminal screens visible?
		Are printouts or photocopies left in plain view?
		Are security awareness posters in evidence?
		Do doors or files have locks?
		Are analog or digital combination locks employed?
		Can an employee tell you the approximate date of the last security briefing conducted in the area?
		Is data encryption employed in the area? Are senstive work documents disposed of securely (shredded, et al)?

REMARKS:

SPECIAL INSTRUCTIONS FOR SECURE AREAS

If the area under observation is a secure area, obtain the name of the security administrator or officer responsible for the area. Set up an appointment to meet with him/her. Develop a suitable strategy for providing secure offsite storage, for accommodating special security requirements in system and network recovery plans, and for accessing locked areas in the event of an emergency.

**NAME OF SECURITY OFFICER/
ADMINISTRATOR:** _____

TELEPHONE AND EXT: _____

DATE/TIME OF MEETING: _____

Figure 6.4 Disaster prevention inspection worksheet. (Continued)

Similarly, local officials may have historical experience to offer regarding natural hazards in and around the company site. For example, they may be able to impart information about flooding areas in the vicinity which may impair utilities or roads or render company facilities uninhabitable. These threat potentials are also worth knowing about to ensure plan adequacy.

System Areas and Switch Rooms

Of key interest to traditional disaster recovery planning is the exposure of the corporate data center and telecommunications switch room to disaster potentials. This remains an important focus of company-wide contingency planning due both to the dependence of the company on automation resources and the increased susceptibility of sensitive electronic equipment to otherwise tolerable environmental factors.

Support Subsystems

Support subsystems provide those requirements without which the operation of electronic devices would either be impossible or extremely difficult. Power, water, and environmental controls are all examples of critical support subsystems.

Power

As previously stated, power is a subsystem requirement for data processing and digital communications. The power subsystem is actually comprised of a network of subsystems that needs to be examined and safeguarded by the disaster recovery planner.

Kenneth Brill, an authority on data-center power distribution, divides this power distribution subsystem into eleven components.

1. A utility point-of-service entry provides the connections between the local power utility and the company facility. The power from the utility is stepped down to internal distribution voltages, then building switch gear channels power through building circuits. Planners need to know how power is being supplied to the building, who owns and maintains the switching gear, and what alternatives exist for supplying power in the event of an interruption at this point.

2. Lightning protection typically consists of arresters located at the point-of-service entry that are responsible for dissipating lightning energy before it causes harm to the facility or powered devices. Planners should discover the type and capacity of arresters and obtain the results of the last inspection of this equipment by utility company personnel.

3. Critical power buses are another subset of the power subsystem for a facility. In a well-prepared facility, there will be three power buses. One is a *raw power bus* that delivers electric power from the utility point-of-service entry throughout the building. The second bus, called an *interruptible bus* is typically a circuit equipped with an emergency power generator to be used if utility power is ever cut off from the building. Third is an *uninterruptible power bus*, which represents the connections between an uninterruptible power supply (UPS) and critical electronic equipment. In the case of a utility outage, and until an emergency generator can be brought on-line, the UPS bus supplies power to operate (or perform an orderly shutdown of) data processing and communications equipment. In normal

operations, the uninterruptible bus is used to condition utility power before it is supplied to electronic gear. Planners should obtain documentation about these buses at their facilities.

4. The UPS itself is another component of the power support subsystem. A UPS is sized and selected to support a "load" represented by the electronic hardware, lighting, and other equipment that it will support in the event of a utility power failure. (Planning activities with regard to UPS systems are discussed in greater detail later in the chapter.)

5. Air conditioning is a prerequisite for proper UPS operation. Often a special chiller unit will be installed in the UPS room to provide optimal operating temperatures and to prevent an automatic temperature-related shutdown of the hardware, which would allow raw, unconditioned power to reach computer and switch equipment. This AC unit should be periodically inspected for proper operation.

6. A frequency converter, a component of a UPS subsystem that is typically positioned near a mainframe that it serves, converts 60Hz utility power to the 415Hz power required by many mainframe processors. The frequency converter thus needs to be inspected periodically for proper operation.

7. UPS system batteries supply power in lieu of utility power for a predetermined period of time (usually long enough for utility power to be restored or for a backup generator to be brought on-line.) The Occupational Safety and Health Administration (OSHA) has developed some fairly lengthy guidelines on the proper placement and care of these batteries. The useful life of batteries is somewhat limited. Thus, planners need to inspect batteries for OSHA compliance and ensure that periodic testing is performed to guarantee that power reserves will be available when they are needed.

8. Emergency generators provide power in the event of a prolonged interruption of utility power. Generators must be sized according to the load they will carry. Related equipment, such as parallel switching gear used to harness the output of two or more generators, must be inspected for proper operation.

9. Testing equipment and load simulators are used to verify that UPS batteries and emergency generators will supply required power when needed. Thus, these off-line testing facilities must be inspected and utilized on a regular schedule.

10. Computer-room power distribution panels or modules deliver conditioned power from the UPS bus to various load devices. This subsystem needs to be checked periodically.

11. Grounding is Brill's final power subsystem component. Expert evaluation should be sought to ensure that grounding is adequate and that employees are not subject to hazardous electrical shocks.

Of course, this description of the "normal" power subsystems in a data center begs one question: What if a facility is without UPS or emergency power generation equipment? If a UPS is not yet installed, the disaster recovery planner should work with data center operations management to purchase or lease one. Few disaster prevention capabilities provide such demonstrable benefits to normal operations as the UPS.

UPS systems may be obtained on a turn-key basis from no fewer than 80 vendors. Typically, the planners can provide a compelling justification for the acquisition by referencing the costs of a single power outage event. Most data-center managers can readily provide this information from their own records of power failures and their consequences; it is likely that the entire user community will back the request without reservation.

To prepare for the acquisition of a UPS, it is valuable to list equipment that will need to be protected by the system. This load equipment may include:

- Lighting in key work areas
- Data-center hardware and certain terminal controllers and networks
- Air conditioning in the UPS room, telephone switch room, and data center
- Company private branch exchange (PBX) or centrex telephone systems
- Electrically powered access control systems (such as ID card readers)
- Electrically powered disaster prevention systems that are not otherwise backed up with their own battery supplies
- Elevators

Figure 6.5 provides a sample worksheet for listing UPS load requirements. Adjacent to each entry, it is useful to enter the manufacturer-specified Volt-Amps or Kilovolt-Amps specification for the unit. (If VA or KVA specifications are not provided, multiply volts by amps to provide this figure.) This is a baseline estimate of the UPS size required. Some engineers recommend obtaining a UPS with an output power rating 30 to 60 percent in excess of this baseline to accommodate the starting current requirements of certain types of devices. A UPS vendor's technical expert will be in the best position to advise on additional startup amps requirements.

Once a load estimate has been developed, selecting a UPS requires that the planner understand the differences between UPS systems. Most UPS systems are of two types: in-line or stand-by. An *in-line* UPS operates constantly within the raw power bus, converting power for storage in its batteries and inverting battery power from DC to AC to provide a clean, continuous supply for load equipment. A *stand-by* UPS switches in to replace utility power when a sudden drop is sensed. This switching process takes only milliseconds; some electronic equipment is relatively insensitive to the transition. However, the switching interval can pose tremendous difficulties for certain types of disk drives, causing data loss or head crashing when the changeover occurs. A vendor engineer will be able to advise the planner about the options and their consequences.

UPS SIZING WORKSHEET

INSTRUCTIONS: List all equipment to be supported by an uninterruptible power supply. Provide the manufacturer-specified Kva requirements for each item. Attach additional pages as needed.

EQUIPMENT	KVA REQUIREMENTS
TOTAL KVA:	

Figure 6.5 UPS sizing worksheet.

Water

With certain models of computer mainframes (and for certain types of air conditioning plants), water can be as vital to operations as electricity. If critical systems hardware at the planner's company is dependent on water, the planner should obtain mechanical drawings of facility plumbing, as well as documentation on public water delivery networks.

The options for replacing public water supplies in the event of an outage are few. A rooftop or freestanding water reservoir may provide a temporary supply, while a private well and pumping equipment may provide a more reliable alternative water source. Some utilities will also construct dedicated water supply lines that bypass local water mains, but only at substantial cost.

If water is an operational requirement at the planner's facility, alternative supply options should be considered. However, it may come as no surprise that a water main break that results in a prolonged outage is usually met by the relocation of data-center operations to an alternative site.

Environmental Controls

Climate and air purity controls are often overlooked in discussions of disaster prevention, despite their association with other types of disaster potentials. Nearly every planning methodologist acknowledges the need for heating and air conditioning controls that maintain the temperature and humidity levels within the data center within manufacturer-specified tolerances. However, very few treat the subject of environmental contamination, which is increasingly pointed to as a source of equipment failures and even fires. Figure 6.6 lists some contaminant types and their damage potentials.

TYPE	DAMAGE POTENTIALS
METALLICS	ELECTRICALLY CONDUCTIVE, MAGNETICALLY ATTRACTED TO MICROELECTRONIC CIRCUITS
CARBONACEOUS PARTICULATE	MOISTURE ABSORBENT, ELECTRICALLY CONDUCTIVE, COMBUSTIBLE
SYNTHETIC FIBROUS PARTICULATE	LOW MELTING POINT, MOISTURE ABSORBENT, SOME TYPES ARE COMBUSTIBLE, ELECTRICALLY CONDUCTIVE
CEMENT DUST/ CHRYSTALLINE PARTICULATE	MAY BE PROPELLED AT HIGH SPEED INTO FLOOR-COOLED COMPONENTS, CLOG DISK MEDIA FILTERS, CAUSE OVERHEATING, PLATTER WARBLE, AND HEAD CRASHES

Figure 6.6 Environmental contamination hazards.

Ensuring that climate and humidity controls are uninterrupted by power loss is accomplished by including these devices in an uninterruptible power bus. Equipment failures can only be prevented through a schedule of inspection and maintenance.

To control environmental contamination levels in areas with sensitive electronic equipment, however, requires considerably more effort. To add contamination control to the preventive systems recommended by the planner, first the planner must identify what contaminants are present in the environment, at what levels, and from what sources.

Currently, VICON™ detectors are the most reliable means for determining contamination levels by type of contaminant. These detectors, available from CCI Technologies (Miami, Florida), can be strategically placed within the data center subfloor plenum and around equipment to take samples from continuous airflow over time. Various media are used to collect contaminants of different types within the detector housing. Subsequent analysis and averaging of the media contamination levels can provide a fairly comprehensive picture of environmental contamination in the facility.

Knowing the types and levels of contaminant particles present in the airflow can provide clues about the origin of contamination so that steps can be taken to eliminate certain sources. Figure 6.7 offers an overview of typical contamination sources.

The current trend toward *dark room* operations in the data center is motivated partly by a desire to reduce the environmental contaminants introduced into the equipment room by human traffic. Human skin, hair, clothing fibers, and cigarette smoke, as well as the presence of printers, printer paper, and wastebins in the computer room are all known contributors to carbonaceous, organic, and synthetic fibrous contamination. These particles have a tendency to *plate out* on equipment surfaces, including circuit boards and filters. Since nearly all particulate can conduct electricity due to moisture absorbency, the results of contamination may include short circuits and even fire. Moreover, the blockage of filters, and the penetration of filter media by microscopic contaminants can lead to the overheating and warbling of disk media or sufficient contaminant buildup on media surfaces to result in read-write errors and ultimately head crashes.

Figure 6.8 summarizes some additional steps that can be taken to reduce contamination levels in the data center or telecommunications switch room. It is not necessary to establish a *clean room* environment, of course, but these steps will reduce contaminants to a less harmful level. If read-write errors or head crashes are already chronic in the data center, planners should explore whether contamination may be the cause.

Figure 6.7 Common sources of contamination.

1. Verify that the tiles in the computer room floor are not worn and that the subfloor is properly sealed. When work is performed in the subfloor plenum, ensure that tiles are replaced in the same position as they were removed. Wear of tiles and concrete surfaces of the subfloor are known contamination sources.

2. Inspect air conditioning plants for wear that may release metallic particles into the environment. Also, ensure that AC is not drawing in unfiltered air from outside the facility, as this may introduce urban pollution to the data center or equipment room environment. Check the American Society of Heating, Refrigeration, and Air Conditioning Engineers (ASHRAE) ratings of filters used in the main air conditioning plant and with all spot coolers to ensure that optimum efficiency filters are used.

3. Establish and enforce a no-smoking policy in the equipment room.

4. Use channel extension or other technologies to position laser and impact printers, and their related supplies (paper, burster/decollators, etc.) outside of the data center. This will reduce both human traffic and levels of carbonaceous contaminants (paper dust, laser printer toner) in the equipment room.

5. Remove space heaters with nicron heating elements, and all ion-generating air purification equipment from the equipment room. Both can charge airborne particulate so they "seek out" circuit boards.

6. Ensure that all equipment is equipped with high quality filters, providing filtration at micronic rating levels. Inspect and replace these filters regularly.

7. Ensure that positive air pressure exists in the equipment room (air should blow outwards from the center whenever a door is opened. If this is not the case, consult with an environmental engineer to determine how positive pressure can be established.

8. Establish a regular schedule for environmental testing using a reliable method for sample collection and evaluation (i.e., VICONTM technology). Use the results of tests to identify any contamination sources that may have been overlooked.

9. Contract for competent computer room maintenance and cleaning at regularly scheduled intervals. Cleaning companies should use filtered vacuums with brushless rotors, non-ionic cleaning solutions, and should never wax computer room floors.

10. If contamination problems remain chronic, explore the possibility of air purity maintenance through environmental filtration equipment.

Figure 6.8 Checklist for reducing environmental contamination.

Additional information on environmental contamination is available from the General Services Administration in the form of Federal Air Standard 209B and from studies published by the American Society of Heating, Refrigeration and Air Conditioning Engineers (ASHRAE) in Atlanta, Georgia.

Water Detection and Alarm

Statistically, flooding is the number one disaster potential in data centers. Flooding may result from a variety of causes including weather events, structural faults in facilities, plumbing leakage, leaks from air conditioning plants and mainframe water cooling systems, and faulty sprinkler systems. Another cause of flood damage is firefighting. Together, these causes result in millions of dollars in flood damage claims to commercial and federal insurers annually. (Federal flood insurance is provided through the Office of Insurance Support Services of the Federal Emergency Management Agency to flood risk areas in the United States, since homeowners and businesses in these areas are unable to receive coverages from commercial insurance companies.)

One obvious result of water intrusion into an equipment room filled with electrically powered equipment is short-circuiting. However, even with equipment powered down prior to water intrusion, damage to sensitive electronic circuit boards and disk media is often beyond repair, despite the claims of some hardware vendors to the contrary.

Vendors of chemical fire suppression systems often cite a study conducted by the Air Force Engineering and Services Laboratory at Tyndall Air Force Base, Florida, (Document ESL-TR-B2-28, 1982), which cites the harmful consequences of water staining, to make a case against water sprinkler systems in computer and switch rooms. Even if equipment has been covered with tarpaulins and is allowed time to dry completely (one vendor suggests using a hair dryer to speed this slow evaporation process along), the systems and networks operated on the hardware, and the data created, managed, and distributed with the systems and networks, will have questionable integrity.

Of course, water intrusion does not necessarily occur all at once. Instead, small leaks over time may create disastrous results. To prevent this possibility, planners should consider the installation of a water detection and alarm capability that will notify data center and switch room personnel to the presence of water or other liquids before they can reach and harm electrical devices.

There are a number of detection systems available. Most utilize an array of sensors which are placed strategically within a protected area. For example, sensors may be placed near walls that house facility plumbing, or they may be placed along water conduits for mainframe water cooling plants, or they may be placed near air conditioning access ducts, and so forth. When the sensors are exposed to excessive moisture conditions, they respond with an alarm to signal personnel about a flooding risk. More intelligent models of these sensor arrays provide the means for signalling the specific location of a leak through some sort of sensor numbering system and alarm grid.

It should be understood that flood detectors will provide no protection from massive flooding due to weather or other factors, but they will help to identify minor leaks so appropriate responses can be made before major downtime and damage result. Thus, planners, whose facilities utilize water-cooled hardware or are designed with restrooms or water piping abutting equipment room walls, should explore a water detection and alarm subsystem for equipment room protection. Those whose facility design poses none of these hazards should consider the need for water detection in the context of other prevention system expenditures.

Fire Detection and Suppression

Fire is often the first disaster potential business managers envision when they hear the term "disaster recovery planning." The threat to human life, as well as to business operations, posed by fire lends this disaster potential a kind of mystique that such factors as water intrusion or environmental contamination or power failure may lack. Still, for all of the discussions surrounding the fire threat, it is flooding and not fires that is the most frequent cause of disasters and environmental contamination, not fires, that results in the greatest amount of equipment downtime.

This is not to understate the importance of fire protection. According to the National Fire Protection Association (NFPA), equipment room fires average about 90 per year, with the average cost of a fire in these areas estimated at approximately $3 million. Some other interesting facts about fire are:

- The temperature at which fires begin to damage computer equipment is 175 degrees Fahrenheit (not the 451 degrees at which paper burns)—and this is 50 degrees more than required to melt tape and disk media.
- Smoke generated by even a small, quickly contained fire introduces tremendous amounts of particulate contamination into the data center that can result in equipment failure or subsequent combustion.
- Water, in combination with burning cable sheathing material (PVC is generally used to sheath computer cables), produces a gaseous mixture of phosgene, hydrogen chloride, and hydrogen cyanide—a combination of acids and deadly fumes that can harm equipment and kill personnel.
- Burning power transformers may release PCBs into the atmosphere that will render a data center uninhabitable for a prolonged period of time due to long-term contamination and health risks.

Fortunately, the risk of fire can be reduced through a time-tested combination of procedural and mechanical systems. Figure 6.9 provides a checklist for fire prevention.

Item five in the checklist, arranging a facility inspection by local fire protection officials, is a key activity. Firefighting professionals bring a learned eye to the protection requirements of a facility, as well as a knowledgeable perspective on the techniques that will be used to suppress fires that cannot be prevented or suppressed by internal systems and methods. They may be able to offer advice on the effectiveness of a decision to purchase chemical fire suppression systems based on the facility layout and equipment room structural considerations. If the room will not sustain an "envelope" of fire suppressing gas— that is, if its design cannot contain the necessary concentration of the suppression agent—the purchase may be unwarranted without structural modifications.

NFPA guidelines, referenced in item six of the checklist, are among the most reliable for fire prevention as they are based on the experience of firefighters and fire protection officials throughout the country. NFPA has developed construction specifications for fire prevention in equipment rooms and in media storage areas that are well worth the planner's time to read and consider when planning for company facilities.

1. Establish a no-smoking policy.

2. Regularly inspect facilities and remove combustibles from the proximity of potential igniters.

3. Identify fire-related health risks—certain types of synthetic carpets and upholsteries, plastics, PVC cable sheathing, etc.—and replace them with less hazardous equivalents.

4. Inspect all electrical devices, extension cords, etc., for UL labels.

5. Contact local fire protection officials and arrange for an inspection.

6. Contact the NFPA in Quincy, Massachusetts to obtain facility fire protection guidelines and other related publications. Identify the fire ratings of equipment room walls, ceilings, and floors.

7. Ensure fire exits are unobstructed and clearly marked. Identify fire and smoke detectors and test for proper operation.

8. Conduct regular fire evacuation drills and initiate a company-wide disaster prevention awareness program.

9. Locate fire suppression capabilities such as standpipe/hose assemblies, sprinkler systems, and wall-mounted extinguishers. Check for current inspection and test labels.

10. Contract for chemical fire suppression in equipment rooms.

Figure 6.9 Fire prevention checklist.

Physical Access Controls

Destruction or damage of automation resources may result from the deliberate or inadvertent actions of persons who have access to equipment rooms. This is another disaster potential that needs to be considered by disaster recovery planners.

Planners need to assess the current level of physical security in equipment rooms. In addition, planners must determine whether physical security provisions might impede the evacuation, recovery, or restoration activities that occur in a crisis situation and its aftermath.

Access controls aim at three goals: to prevent theft, to prevent damage, and to prevent unauthorized disclosure or use of company electronics and data. Preventing theft of a mainframe or large PBX is simplified by the physical size and weight of these objects. However, smaller hardware items, such as PCs, terminals, desktop printers, and so forth, as well as removable electronic media, blank check stock for printing payroll or accounting checks, check logs, and other items may be prey for intruders or unscrupulous employees.

If the truth be known, it is quite easy to damage practically any electronic device. Fires can be set, cables cut, water supplies shut off, and power interrupted. Moreover, media containing important company data can be damaged unintentionally or deliberately by placing the media on which it is stored in proximity to a magnetic device, such as a paper clip holder. Thus, in addition to keeping out intruders, physical access controls also aim to restrict access to equipment rooms to those who have sufficient knowledge about electronics to prevent them from making obvious mistakes that could sabotage equipment or data.

Finally, access controls are often employed to preserve corporate trade secrets or sensitive information about company operations or personnel. Even photographs of proprietary equipment configurations may represent an exposure of company secrets that could harm its ability to compete in the free marketplace.

Thus, there are a number of valid reasons to undertake physical security planning and a number of methods have been developed to accomplish physical security goals. These methods may be categorized into three types:

1. **Authentication Controls**—Authentication controls are used to verify that persons seeking entry to a sensitive area are authorized to do so. Authentication may be accomplished through some sort of identification tag that is evaluated by a human or electronic authenticator before access to the equipment room is permitted. The simplest form of authentication control is a lock to which only authorized individuals have a combination or key.

2. **Activity Monitors and Alarms**—Another concern of physical security is the activities of persons admitted into a sensitive area. These activity controls monitor the locations of individuals (e.g., logged access systems), the movements of individuals in sensitive areas (e.g., videotaping systems), and/or alarm when improper activities are detected (e.g., tamper sensors, intrusion alarms, etc.).

3. **Exit Controls**—Exit controls are used to control the removal of sensitive data or equipment from the equipment room. Techniques may involve manual inspections of briefcases and other containers, the use of secure disposal systems (e.g., shredders), electronic controls on voice and data communications lines, or devices that prevent the duplication of data on disks.

Depending on a company's perceived need for security, these controls may be fairly complicated or they may be relatively lax. For the purposes of disaster recovery planning, these physical access control measures must be documented and evaluated for adequacy. Where controls are inadequate, alternatives and enhancements should be recommended for installation. Figure 6.10 provides a checklist for physical access control planning.

Authentication controls must also be examined as possible impediments to disaster recovery. For example, in restricting access to an equipment room through the use of code keys and locks, it may be difficult to enter the room promptly and power down equipment in the event of a sudden water pipe burst. Moreover, security measures may impede efforts to evacuate check logs and other operations records that are required for recovery but kept under lock and key in normal operations.

---

1. Identify authentication devices used to control access to the data center or telephone equipment room. Assess controls for completeness and suggest additional security as needed.

2. Identify electronic and manual locks on doors and windows. Ensure that exit is not restricted. Suggest additional locks as needed.

3. Obtain copies of keys or keycodes that may be used to surmount physical access controls in an emergency.

4. Identify sensitive materials storage access controls. Obtain keys or codes required to evacuate sensitive materials in an emergency. Suggest additional access controls as needed.

5. Establish blind drops or other methods to ensure that access control information is not revealed except in an emergency.

6. Establish a physical security awareness program for equipment room personnel.

Figure 6.10 Checklist for physical access controls.

Activity monitors and exit controls may also provide security and, at the same time, impair recovery. In some facilities, emergency exits are locked to prevent unauthorized removal of sensitive materials. This may create hazards for employees who need to exit the facility quickly in the case of a fire. The same situation may result if systems that detect unauthorized activity also cause automatic locks to activate, preventing access to or exit from sensitive areas. (In some facilities that install chemical fire suppression systems, these automatically lock and seal a protected area within minutes of hazard detection so that an "envelope" is created before the suppressive agent is "dumped." Despite vendor claims that the fire-suppressing gas is not hazardous to humans, being locked in a suppressant-filled room may offer little consolation if the rest of the building is burning.)

It is important to obtain combinations for electronic and mechanical locks, as well as copies of any keys; they should be available for use by personnel who may be called upon to act first in an emergency. More than a few security professionals bristle at the thought of this breach of security. However, one method for satisfying both the needs of the security officer and those of the disaster recovery planner is to keep this codes and keys in a sealed envelope, only to be opened in the event of an emergency. All operations personnel need to be made aware of this so-called *blind drop* and when and how it is to be used. Also, they need to be aware that the envelope will be regularly inspected for tampering or unauthorized disclosure and of the ramifications for security violations.

Electronic Access Controls

In addition to physical access controls, many firms employ a range of software and hardware controls to prevent unauthorized access to company systems and networks. As with physical access controls, the intent is to lock vandals, hackers, and other computer criminals out of corporate databases and networks. Two major categories of electronic protection exist:

1. **Port Protection**—Port protection aims at protecting unauthorized entry into systems and networks by securing the access ports by which such entry is accomplished. A combination of hardware and software controls are often used to protect communications ports, including call-back modems, silent modems, and hardware code validation. Other measures, such as data encryption devices, prevent the use of data by persons who are able to defeat port protection and password protection measures, or who successfully "tap" into a valid communications session.

2. **Password Protection**—Perhaps the most common form of network and system access control is password authentication. Legitimate users of the corporate telephone system or corporate computer applications are assigned unique passwords. When they seek access to a system or network service, they must enter the password, which is validated by software on the computer or PBX. If the password checks, access is granted to programs and data for which the user is authorized according to a *user profile* associated with the password. More capable password protection strategies further provide means for monitoring the activities of a user once access has been granted. Efforts by the user to access unauthorized programs or files, to perform file deletions, or to introduce new programs or modify existing programs (i.e., in the case of computer viruses), are logged and the user is warned that the activity is not within the scope of his authorized activities or is disconnected.

As in the case of physical access controls, electronic access controls are of interest to the disaster recovery planner both from the perspective of their adequacy as a hedge against security-related downtime and as a potential obstacle to efficient recovery. As a first step, planners should work with information security officers and auditors to identify existing controls. Then, any additional measures that may be needed should be documented and proposed to management.

Planners also need to pay particular attention to potential obstacles to disaster recovery posed by electronic access controls. For example, if hardware identification methods are used to validate terminals whose users are requesting access to company systems, this security measure may create problems for restoring terminal networks if new terminals are used. Dial-back modems must be reprogrammed so that they return verification calls to new telephone numbers in the event that the original user location is no more. The list goes on and on.

Of course, electronic security will also need to be provided at the recovery site. Thus, efforts should be made to ensure that software security is portable to other environments where recovery will occur.

User Work Environment

In general, the prevention capabilities previously described are also required in user work areas. Arguably, fire prevention, theft prevention, and access controls are even more important in the user environment since this is where disasters are more likely to originate (or thefts more likely to occur) than in data centers or telephone switch rooms.

DISASTER PREVENTION CHECKLIST

WORK AREA: _____
NUMBER OF PERSONNEL: _____

AUTOMATION RESOURCES: _____
_____
_____

MANAGER: _____ _____
SECURITY ADMINISTRATOR: _____ _____

PHONE & EXT

☐ Blueprints and floorplans for work areas have been provided to the disaster recovery coordinator and are current.

☐ Emergency exits are clearly marked, readily accessible, and employees have been informed of their locations and use.

☐ A disaster recovery/security awareness training program has been developed and administered to all work unit employees.

☐ Environmental testing has been conducted and particulate counts have been found to be within acceptable limits. (Identify testing method, date of test, and count: _____)

☐ Power and surge protection requirements have been identified and adequate uninterruptible power supplys have been acquired.

☐ Physical security measures have been identified and lock combinations, passwords, duplicated keys, etc. have been made available to the disaster recovery coordinator.

☐ A fire hazard survey has been conducted and identified hazards resolved.

☐ Software and hardware resources have been identified and inventory lists have been provided to the disaster recovery coordinator.

☐ Supply requirements have been identified and a supply list has been provided to the disaster recovery coordinator, together with a list of supplier contacts.

☐ Records requirements have been identified and a records list has been provided to the disaster recovery coordinator.

☐ Procedures have been developed to duplicate records and data for off-site storage and to review off-site storage at periodic intervals to identify and cull stored recoverds and data that no longer need to be stored.

Page 1 of 2

Figure 6.11 Disaster prevention checklist.

☐ Fire alarm and suppression systems have been tested and verified for proper operation. (Provide date of last test/inspection: _____)

☐ A disaster recovery pont-of-contact (and backup) has been appointed to serve as the reviewer of the disaster recovery plan, as it relates to this work unit, to participate in testing, and to request changes to keep the plan up-to-date. This person is:

PRIMARY CONTACT: _____
LOCATION: _____
OFFICE PHONE/EXTENSION: _____
HOME PHONE: _____

SECONDARY CONTACT: _____
LOCATION: _____
OFFICE PHONE/EXTENSION: _____
HOME PHONE: _____

☐ A notification list has been developed and copies have been provided to the disaster recovery coordinator and the primary and secondary points-of-contact containing persons to be contacted in the event of a catastrophe.

☐ This work unit is participating in a company-wide program of off-site storage. Our regular pickups are provided on (describe pickup arrangements: _____ .)

☐ Within the work area, (give name) _____, is authorized to request emergency deliveries from the off-site storage vendor. Our off-site storage artons bear the code _____ to distinguish them from those of other work areas. The name and phone of the off-site storage vendor is:

VENDOR & CONTACT: _____
LOCATION: _____
NORMAL BUSINESS PHONE: _____
EMERGENCY PHONE: _____

Figure 6.11 Disaster prevention checklist. (Continued)

Figure 6.11 provides a straightforward worksheet and checklist for use by work unit managers in assessing and documenting their own disaster prevention capabilities and general emergency preparedness. (Note that some items on the form pertain to off-site records storage and their use will become clearer after Chapter 6.)

At the top of the form, the work unit manager is asked to indicate the automation resources located in the work area, among other things. Here, such items as PCs, LANs, and departmental minicomputers and PBX equipment should be listed—that is, devices that are located within the work unit itself. These resources represent smaller versions of the large systems equipment rooms previously cited. However, if the automation resources are mission-critical to the work unit and the business functions it supports, these systems and networks too must be considered in a disaster prevention capability.

Decentralized Computing

By the end of the 1980s, decentralized computing had become a fact of life in corporate America. This trend was fostered by the increasing power and capability of small processors in the form of minicomputers, microcomputers, and local area networks (LANs).

From the beginning, advocates of centralized control of automation resources argued that distributing control among departmental lines would result in chaos. Indeed, one hazard of decentralizing corporate computing was the increased vulnerability of information systems security and data integrity. At a program development level, there would be no standardization among applications developed by numerous independent work units—in terms of programming languages, system architectures, or file structures. Hence, the result would be numerous, disparate applications without file compatibility: a nightmare for integration.

To some extent, the fears of the detractors of decentralized computing have been realized. Programs or database applications are often developed within work units without regard to company programming standards. Inadequate documentation and employee turnover have often been cited as a source of operational malaise affecting the efficiency of the work unit as a whole.

What this disparity means to disaster recovery planning is more work. Simply, systems—whatever their size—that provide critical functions for the company must be restored. If different departments utilize entirely different and incompatible operating systems and hardware, provisions must be made to provide suitable replacement systems in the event of a disaster.

Disaster recovery planners thus need to attempt to influence the acquisition and implementation of departmental systems. The effort need not take the form of a general condemnation of decentralized computing, which would be tilting at windmills. Rather, planners should seek to encourage the design and acquisition of decentralized systems that can share files or that can be ported to centralized processing platforms in the event of disaster. The latter is at the heart of a minimal equipment configuration recovery strategy that will be addressed in greater detail in Chapter 8.

This effort falls under the category of disaster prevention in that it seeks to prevent a prolonged outage based on difficulties in obtaining suitable replacement hardware. The concept can also be extended to any device for which a replacement would be difficult to find. For example, if an application is being supported on an outdated hardware platform because the application works well and has never been converted for use on more modern equipment, the disaster recovery planner should ascertain whether disaster recovery considerations mandate that the system be converted into a more recoverable design.

Disaster recovery planners should attempt to document disaster recovery considerations in systems development for distribution to work unit managers and system designers within work units. Clearly explaining the relevance of company programming standards to recovery may aid in promoting the adherence of departmental programmers to the standards. If this reasonable approach fails, the nonstandard systems and applications should be cited in reports to senior management and in plan tests.

Microcomputers

Microcomputers and related workstations have special disaster prevention requirements based on their mode of use. Typically, a single user will manage the applications and data on a microcomputer. The user must, therefore, become an active participant in ensuring the safety and recoverability of the system.

Most important among all disaster prevention strategies for microcomputer-based systems is a regular program of data backup. The applications and data used on micros should be backed up to removable media at regular intervals. The reasons for this are several:

- Microcomputer components generally have much shorter mean-time-to-repair (MTTR) and mean-time-between-failure (MTBF) intervals than do large system platforms. Hard disk failures occur, on average, every 20 to 30 thousand hours (roughly every two to four years). Improper use and environmental factors may reduce this life expectancy substantially.
- Often microcomputers are powered from the raw power bus used to distribute electricity throughout the building. Surges, spikes, blackouts, brownouts, and line noise are all potential causes of component failure, system failure, and read-write errors.
- Environmental contamination levels in user work areas are nearly impossible to control without establishing *clean rooms*—an overly expensive and impractical alternative in most companies. Thus, contaminant build-up may impair the operation of microcomputers over time.
- Open work areas provide virtually no security for microcomputers. While password controls are available for micro systems, they are rarely implemented. Thus, microcomputers offer a "social" computing environment that exposes them to viruses and other software-based attacks.
- Data center fires account for only a small percentage of all fires affecting businesses every day. Most begin in user work areas. Since non-water fire-

suppression agents would be impractical in a facility that does not provide an airtight envelope, microcomputers are likely to be saturated by sprinkler systems or fire department hoses in the event of fire. Subsequent use of the systems would prove unreliable.

Thus, the best strategy for preventing disasters involving microcomputers is to regularly back up the applications and data on these systems and remove backups to a secure location at an alternate site. In this way, the loss of a comparatively inexpensive and often easily replaced piece of hardware need not result in a disaster.

Beyond this strategy, other preventive measures may be added if they are justified by the criticality of micro-based applications. Small UPS units and surge protectors are available to supply clean, uninterruptible power to micros. Plastic covers are available to shield these devices from water damage if sprinklers are activated or roof leaks develop. As previously stated, software products are available both to secure micros from unauthorized use and to check for virus infections, prevent hard disk formatting, and otherwise foil certain types of system sabotage. Finally, security measures—from physical access controls to theft prevention—may be implemented in any work area that needs them.

Planners should examine all micros, identify their purpose and users, and account for them in their disaster recovery plans. Also, users must be made aware of their responsibilities for safeguarding their automation resources from loss or damage. A formal off-site backup strategy should be articulated and enforced.

LANs

One method for deriving some of the disaster recovery benefits of centralized computing in the decentralized environment of end-user microcomputing is to connect the workstations into a local area network. Most LAN operating systems provide optional security capabilities to provide password protection capabilities. Moreover, with a LAN, such administrative functions such as LAN backup may be accomplished automatically from a central location.

In addition, LAN file servers need not be co-located to networked workstations. In some implementations, the fileserver may be located within the data center where it is afforded a more secure and less disaster-prone operating environment. In other topologies, the LAN server may actually be the corporate mainframe itself.

In a fileserver-based LAN (there are distributed server LANs in which every workstation shares control and administration of LAN functions), the server should be accorded the disaster prevention capabilities that accrue to its function as the central repository of data files and applications. UPS protection, environmental controls, and fire and water protection monies are best spent safeguarding the LAN's data repository, the server, from harm.

Other Equipment

In addition to processors, other equipment may require special disaster prevention measures. Small telecommunications switches, channel extension hardware used to connect work unit terminals to a distant mainframe, and computer peripherals whose operation is critical to system performance (i.e., modems, multiplexers, device controllers, item processors, etc.) should all be scrutinized for their protective and preventive requirements.

IDENTIFYING VENDORS

Having identified appropriate disaster prevention technologies, next it is necessary to find vendors of these technologies and to compare products so a selection can be made. Many companies have a purchasing department to handle the acquisition of products once bid specifications have been released and bids received and evaluated. Planners should seek to work within corporate purchasing procedures when selecting and acquiring products.

Figure 6.12 is a sample specification for disaster prevention capabilities that might be sought by a company. The form is divided into six sections—five pertaining to equipment rooms, one to user work areas. The five equipment room sections cover requirements in the areas of power backup, water detection, fire prevention and detection, environmental maintenance, and physical access controls. The user work area section is much more generic and may be used to specify a diversity of preventive and protective capabilities.

The form, once completed, is intended to be forwarded to company purchasing. If no such department exists, planners may use the form as an internal worksheet, understanding that individual specifications will need to be drawn up for the requirements identified in each section so they may be disseminated to specific vendors.

Initial Contacts

How can planners locate vendors for the products or services they are seeking? Once again, the best sources of information are disaster recovery periodicals, disaster recovery professional organization vendor directories, and referrals from business acquaintances.

A rule of thumb adhered to by many companies is to solicit at least three bids from qualified vendors. Doing so provides a better picture of current market pricing for the services or products sought. To summarize the bids of the three vendors, a worksheet similar to Figure 6.13 may be employed.

Planners need to be sure that all specified requirements are provided by each bid. Moreover, vendors should be contacted directly to determine whether all products and services will be provided by the vendor or whether portions will be supplied via subcontractor arrangements. Basically, every effort needs to be made to guarantee that products and services of comparable capability and worth are being bid. A bid that has an associated cost which is substantially less than other bids may be an indicator that the vendor is not bidding comparable products or services. Some communication with vendors will be required to resolve discrepancies.

DISASTER PREVENTION BID SPECIFICATION WORKSHEET

POWER (EQUIPMENT ROOMS)

REMARKS

UPS KVA REQUIREMENT? _____

IN-LINE OR STAND-BY? _____

AC BY UPS CONTRACTOR OR OTHER? _____

LEASE OR BUY? _____

NUMBER OF ALARM STATIONS? _____

NUMBER OF EMERGENCY CUTOFF STATIONS? _____

IS SERVICE-ENTRY CUSTOMER CONTROLLED OR THIRD PARTY? _____

Enter contacts, business names, and phones for four prospective UPS contractors.

WATER DETECTION

REMARKS

ARE SENSORS TO BE POSITIONED FOR AREA OR SPOT COVERAGE? _____

INDICATE # DETECTORS (SPOT) OR SQ FT FOR AREA TO BE COVERED? _____

IS LOCATION DATA DESIRED FOR ALARMS? _____

NUMBER OF ALARMS? _____

NUMBER OF CUTOFFS? _____

Enter contacts, business names, and phones for four prospective water detection system contractors.

Page 1 of 3

Figure 6.12 Disaster prevention bid specification worksheet.

FIRE PREVENTION (EQUIPMENT ROOMS)

REMARKS

HALON OR WATER?
AUTO OR MANUAL?
NUMBER OF DETECTORS?
SQ FT/CUBIC FT COVERAGE?
NUMBER OF ALARM STATIONS?
NUMBER OF EMERGENCY
CUTOFF STATIONS?
IS PROTECTED AREA SEALED
WITH POSITIVE PRESSURE?

Enter contacts, business names, and phones for four prospective fire protection system contractors.

ENVIRONMENTAL MAINTENANCE (EQUIPMENT ROOMS)

REMARKS

TEST CONTAMINANT LEVELS?
CLEAN & SEAL SUBFLOOR?
CLEAN RAISED FLOOR?
CLEAN EQUIPMENT SURFACES?
TESTING FREQUENCY DESIRED?
CLEANING FREQUENCY
DESIRED?
ON-GOING AIR
PURIFICATION SOUGHT?

Enter contacts, business names, and phones for four environmental maintenance contractors.

Figure 6.12 Disaster prevention bid specification worksheet. (Continued)

PHYSICAL ACCESS CONTROLS (EQUIPMENT ROOMS)

Describe controls that are being sought.

Enter contacts, business names, and phones for four security system contractors.

PREVENTIVE SYSTEMS (WORK AREAS)

Describe systems that are being sought in detail.

EXAMPLE: 10 SMOKE DETECTORS, wall mounted, photoelectric operation; 5
MICROCOMPUTER UPS Systems, 1500watts, 30 minute backup power capacity, 1
millisecond switching; 2 FIREPROOF CABINETS, legal size, 6 hour fire rating, key locks

Page 3 of 3

Figure 6.12 Disaster prevention bid specification worksheet. (Continued)

PREVENTIVE SYSTEMS BID EVALUATION WORKSHEET

PRODUCT OR SERVICE BID: _____

OPENING DATE: _____

CLOSING DATE: _____

	VENDOR 1	VENDOR 2	VENDOR 3
1. Company Name, Address, Contact Name, and Phone:			
2. Detailed description of product or service bid:			
3. Bid Price			

Figure 6.13 Preventive systems bid evaluation worksheet.

Evaluating Offerings

Once bids have been received and checked for discrepancies, planners need to evaluate the bids for their appropriateness and vendors for their reliability. This may involve finding answers to a number of questions summarized in Figure 6.14.

Planners will want to verify the "pedigrees" of the vendors making proposals. It is important to check the history and tenure of the vendor to ensure that the vendor is experienced in the work that is being proposed, that the vendor has a record of successful installations, and that the vendor will continue to service the installation after the sale. It may also be beneficial to know whether the vendor has performed similar installations for clients in the same industry as the planner's company, as this might suggest a greater familiarity on the vendor's part with the special requirements and scheduling peculiarities of the planner's company.

---

1. How long has the vendor worked in this area?

2. How many customers does the vendor have?

3. Has the vendor performed quoted work for or provided quoted products to other companies within the same industry area?

4. Is the vendor licensed and bonded?

5. What certifications (professional or technical associations, federal government approvals, etc.) are held by the vendor?

6. Are vendor personnel specially trained, certified, or licensed?

7. Does the vendor have a good record with the Better Business Bureau?

8. Does the vendor warranty work? What are the provisions of the warranty?

9. Has the vendor provided manufacturer literature on all products that are to be used? Do the products comply with industry standards (i.e., Underwriting Laboratory, National Fire Protection Association)?

10. Has the vendor provided reference client information?

11. Are vendor sales and technical representatives knowledgeable?

12. Do vendor representatives appear sensitive to the operational requirements of the business and have they indicated a willingness to work within schedules determined by these requirements?

13. Does the vendor service what it sells? Is on-site maintenance provided in the contract?

14. Does the vendor provide training for client personnel?

15. Does the vendor provide after-sale support in the form of periodic preventive maintenance or site inspections?

Figure 6.14 Checklist for evaluation preventive systems vendor bids.

In addition to vendor pedigree, it is also important to check vendor certification, licensing, and bonding. Most contractors need to possess state-issued licenses; special technical certifications may be required for electrical, plumbing, and other specialty installations. Finally, bonds often need to be posted in connection with contractor licenses to safeguard against accidental damage to client facilities, and so forth. Find out whether the vendor is bonded and insured and verify all information, including licenses and certifications, with issuing authorities.

The Better Business Bureau and other consumer protection agencies should be consulted as a matter of form to discover whether the vendor has been reported by other customers for poor workmanship or other complaints. If negative reports exist, talk with the vendor to discover the history of the complaint.

Vendors should be able to provide at least three reference accounts that planners can contact to obtain more information about vendor performance. Reference accounts should be representative of the same type of work that is being sought from the vendor by the planner. Figure 6.15 is a worksheet that may be used when collecting information from reference accounts.

If possible and appropriate, see whether the vendor will work with client accounts to accommodate site visits by the planning team. A site visit can complement a reference account interview by affording planners an opportunity to see the vendor's work first hand, and possibly to allow planners a chance to speak casually with their peers in the host firm.

In addition to checking the vendor's background and experience, evaluating the offerings of vendors also requires that planners investigate the quality of products the vendor will be providing. Often, vendors of preventive systems are contractors who will be installing the products of manufacturers with whom they have marketing agreements. Vendors should be able to supply product information on the products they intend to install. This information should include data on product testing and its approvals by applicable agencies, such as OSHA, Underwriting Laboratories, and other standards-setting bodies.

Additionally, the planner should identify what warranties the vendor is making regarding the products to be installed. If repairs or maintenance are needed at a future date, the planner should learn whether these services would be provided by the manufacturer or the vendor and whether the services will be performed at the installation site, at the vendor facility, or at a manufacturer-specified location.

Finally, the planner should record subjective observations about vendor representatives for later reference. The perceived competence of sales representatives and technical consultants may contribute to breaking a tie between two or three competitive bids.

REFERENCE ACCOUNT INTERVIEW WORKSHEET

CONTACT NAME: _____

TELEPHONE: _____

COMPANY: _____

VENDOR: _____

DATE OF SERVICE/INSTALLATION: _____

Yes No   REMARKS

		WOULD YOU USE THIS VENDOR AGAIN?
		DID THE VENDOR WORK AROUND NORMAL OPERATING SCHEDULES?
		DID THE VENDOR PROVIDE COMPETENT AND KNOWLEDGEABLE PERSONNEL TO SUPPORT INSTALLATION?
		WERE BIDS ACCURATE OR WERE ACTUAL COSTS IN EXCESS OF ESTIMATES?
		DID THE VENDOR PROVIDE TRAINING?
		DOES THE VENDOR PLAY AN ON-GOING ROLE IN SUPPORT, TESTING OR MAINTENANCE?
		WAS THE VENDOR'S PRODUCT OR SERVICE COMPETITIVELY PRICED?
		ARE THERE NEW PRODUCTS OR VENDORS AVAILABLE TODAY THAT YOU WOULD RECOMMEND INCLUDING IN A BID EVALUATION EFFORT?
		DO YOU RECOMMEND THIS VENDOR TO BUSINESS ASSOCIATES?

DESCRIBE INSTALLATION OR USE:

Figure 6.15 Reference account interview worksheet.

DOCUMENTING BIDS

Whether the sought-after preventive systems installation services represent significant expenditures, planners should be sure to document bids very closely. Such documentation may be required to cost-justify acquisitions to management, to show to auditors, for use in tax inquiries, or to substantiate decisions if they are challenged. Documentation of all vendor warranties and representations are also valuable if work that is performed or products that are installed do not meet expectations.

All vendor proposals, product literature, reference account interviews, and site visit notes should be retained in a permanent file. Additionally, criteria employed to select a vendor should be documented and retained.

Vendor quotes are generally valid for a fixed period of time—often 30 to 90 days. If quotes are being solicited for a number of services or products that will be presented for management approval all at once, ensure that the expiration dates of the quotes are known so that bid prices can be updated before funding requests are made.

SUMMARY

Disaster prevention systems make an immediate risk-reduction contribution to the company. A growing group of disaster recovery commentators recommend that planners should strive to identify preventive requirements and implement capabilities as soon as possible following project initiation both to provide interim protection and as a hedge against a loss of management support for planning later on.

Checklist for Chapter

1. Inspect work areas using a prevention checklist. Compile results of findings.

2. Identify UPS/power-conditioning requirements. Size UPS systems for work areas.

3. Acquire and install UPS and power-conditioning systems.

4. Identify water requirements. Assess options for alternate water supplies.

5. Identify climate controls in equipment rooms. Assess adequacy.

6. Contract for a data center/equipment room contamination assessment. Take appropriate actions to reduce levels of environmental contamination as needed.

7. Identify water detection requirements. Assess options for water detection and implement selected option.

8. Identify fire detection/suppression capabilities. Assess adequacy and implement new capabilities as needed.

9. Assess physical access controls. Ensure that keys, combinations, passwords, PIN numbers, and other access restrictors are documented and available in the event of a disaster.

10. Identify and solicit bids from vendors for large or costly facility disaster prevention systems. Consult with management and obtain approval for their acquisition. Coordinate installation.

11. Identify insurance premium reductions or other benefits that will accrue to disaster prevention system improvements.

Developing an Off-site Storage Program

OVERVIEW AND PURPOSE

The speedy restoration of mission-critical business functions in the wake of an unplanned interruption is what prevents an interruption from producing disastrous consequences for a business. Such restoration requires up-front planning—disaster recovery planning—which establishes strategies for system, network, and user recovery at alternative locations.

In a nutshell, this is the rationale for undertaking the disaster recovery planning project. The project seeks to identify critical information resources, then to establish logistics and procedures to assure their continued availability throughout the period of a crisis.

An important component of this endeavor is planning for the safe storage and prompt restoration of company data and records, which are at the core of purposeful information processing. In fact, it may be persuasively argued that without a provision for backing up critical data and records, recovery cannot occur. Equipment can be replaced, new facilities can be located and adapted for emergency use, and communications networks may be rebuilt all within hours of a disaster. However, if data and records are lost in a conflagration—and backups of the lost information are unavailable—work cannot proceed, even if all other resource requirements are met.

For this reason, many disaster recovery planners regard off-site storage of critical data and records as a prerequisite for disaster recovery. A few have gone so far as to claim that planning for regular backup and off-site storage of critical data and records comprises 99 percent of a company's emergency preparedness. All other activities of disaster recovery planning are secondary in importance to this planning event.

THE DISASTER RECOVERY PROJECT
OFF-SITE STORAGE PLANNING

MAJOR ACTIVITIES

Identify Existing Storage Plans
Develop Storage Requirements List
Evaluate Vendor Proposals
Develop Backup Schedule
Implement Off-Site Storage Plan

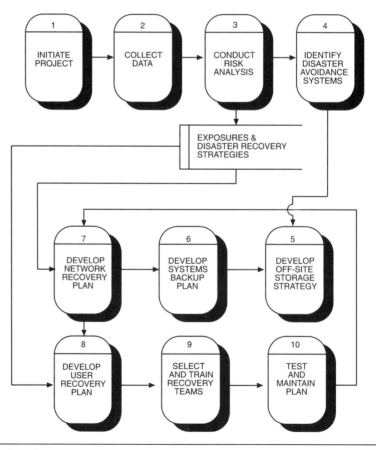

THE DISASTER RECOVERY
PLANNING PROJECT

Figure 7.1 Off-site storage planning phase.

This chapter examines the major activities of off-site storage planning. As enumerated in Figure 7.1, these activities include identifying critical data and records and existing plans for their preservation, developing a comprehensive storage requirements list for the company, identifying suitable vendors for off-site storage and evaluating their proposals, developing a schedule for backup of critical records and data, and implementing an effective off-site storage plan.

Insurance Considerations

Putting aside debates over the relative importance of off-site storage planning and other planning project events, the fact remains that the special significance often accorded to data backup and safe storage is reinforced by both insurance and legal considerations. In each case, the failure to safeguard data from loss creates a financial exposure for the company.

Looking at the issue from the standpoint of insurance, a straightforward case for off-site storage presents itself. While insurance companies do offer media insurance and other special coverages related to data and records recovery, these coverages are limited in what they can do to restore lost data. For example, insurance payouts may be used to purchase new magnetic and optical media to replace damaged media. Moreover, they may be used to reconstruct data that has been lost, either through media salvage and reclamation processes (assuming salvage is possible) or through the use of data-entry services (assuming that primary documentation, originally used to enter data, is available). However, in any case, what is insured is physical media or the costs of salvage efforts and not the data itself. Data, once lost, is gone forever.

Can data itself be insured? The value of data is virtually impossible to determine. Thus, it would be difficult (though some insurers are willing to do so) to purchase insurance in the proper amounts to cover the expenses accrued to data loss. It would be extremely difficult to do so at all cost-effectively!

Legal Considerations

In addition to the business exposures accruing to data or records loss, companies face additional legal exposures based upon their failure to take adequate steps to safeguard data and records that they are required by law to retain. The IRS, the agencies of the Federal Financial Information Examination Council, and several other federal and state government agencies have established records retention and protection requirements with specific fines and penalties for companies that are discovered to be out of compliance.

However, an even greater financial exposure is represented by the civil suits that may be brought by company shareholders in the wake of a disaster. A growing body of law establishes that the directors, officers, and managers of corporations have certain duties to manage and protect their company assets in the best interests of shareholders. These responsible persons have a *duty of trust* (to act prudently in their positions), a *duty of care* (to make reasonable efforts to investigate the condition of the corporation and act to preserve its assets), and a *duty of good faith* (to act in a way that is known to be right). Based upon these duties, the *personal* liability of business leaders is viewed by legal observers as increasing.

The Federal Foreign Corrupt Practices Act of 1977, for example, includes a requirement (in Section 102) for companies to take reasonable steps to assure that accurate books and records are maintained and preserved. This requirement translates to personal liability for corporate officers, internal auditors, comptrollers, treasurers, tax directors, and Management Information Systems directors; it applies to all companies in the United States. Additionally, fines and prison sentences have been established by federal and state governments that punish the failure of a company, its directors, and its officers to preserve records that may be required in the course of an audit or investigation. However, these government-levied punishments are only the tip of an iceberg of possible civil suits based upon a purported breach of duty by company officials.

While Director and Officer (D & O) Insurance exists to protect a company's key employees from lawsuits derived from a purported breach of their duty of trust or care, and while incorporation in Louisiana, Maryland, Nevada, New Jersey, or Virginia further limits the liability of directors and officers of corporations in such lawsuits (due to indemnification rules in the laws governing incorporation in those states), practically no legal protection exists for company officials from a breach of the duty of good faith. Thus, failing to act to do "the right thing" (to comply with the law or to act prudently so as to minimize a known risk or exposure) remains a demonstrable breach of the duty of good faith and exposes company officials to personal liability.

This represents an enormous financial exposure for companies that have failed to develop records preservation (or disaster recovery) plans in the wake of a disaster. In some cases, companies are subject to government fines and penalties for breach of law. However, if it is demonstrated by an irate shareholder that the company officials failed to comply with prudent requirements for preserving the information asset and that this subsequently impaired recovery, lawsuits may follow and large awards may be granted in personal claims against company management.

The simple fact is that the only way to safeguard data from loss (and companies and their officers from lawsuits) is to develop regular procedures for copying data—creating *data backups*—and for removing copies (or originals) to an alternate location for safekeeping. This effort begins by assigning criticality to data and records and developing a suitable backup and storage plan.

ASSIGN CRITICALITY TO DATA AND RECORDS

Which records and data need to be protected? This question presupposes that some records are not as important as others. Clearly, there are differences in the importance of records based on their use or purpose. Some are needed to support day-to-day business activities, others have historical importance and may be subject to special audit or legal retention requirements, and still others may contain sensitive, classified, or trade secret information. These differences need to be understood by the disaster recovery planner in order to develop an appropriate plan for record duplication (backup), storage, and access.

Thus, one of the first steps in off-site storage planning is to identify what types of records and data are being produced by company work units and how the records and data are used. Once this is done, some scheme of ranking is often implemented to assign *relative criticality* to the records and data, indicating which will be included in a program of secure storage and whether they are to be permanently stored or updated on a periodic basis.

Definition of Criticality

Many authors and consultants have advanced taxonomies for ranking the criticality of records and data. Most taxonomies agree on one point: A record or dataset is critical when it is required for the performance of a critical business function. For example, an active dataset that is accessed and used each time a customer order is entered into an automated order entry system would be a critical dataset. In short, any dataset or record that is used in conjunction with an information system or network that has previously been defined as critical shares the classification of that system or network and becomes critical.

Going further, records and data may also be ranked as critical if they are required for use by business functions that support the performance of critical business functions. Just as systems and networks may be defined as critical because they support (provide a necessary input) the operation of other critical systems and networks, so, too, do the records and data of the supporting system share this distinction.

From the standpoint of disaster recovery planning, critical records and data are those that are required for recovery and without which recovery would be impossible. But, is this a comprehensive definition? As previously stated, legal and audit requirements mandate that certain types of records—management reports, accounting data, contracts, the company books—also be protected from loss. Also, proprietary formula, software source code, company-developed processes awaiting patent, and so forth represent an investment of company time and resources or provide a company marketing edge that merit their protection from loss. These records and data are also clearly critical, but are not required to sustain the day-to-day operations of the company.

Thus, the identification of critical records and data may be best made by applying the following criteria:

1. Is the record/data required to perform or provide a critical business function?

2. Is the record/data required to satisfy a legal or audit mandate?

3. Does the record/data comprise proprietary, trade secret, or otherwise sensitive information that would be difficult or expensive to reproduce or reconstruct?

If these questions are kept in mind as records and datasets are reviewed by the disaster recovery planning team, critical records may be distinguished from less critical ones. What is the advantage of such a distinction? A major advantage is cost-savings. According to industry statistics, as many as 75 percent of the files and records retained by company personnel are duplicates of originals, out-of-date, or never referred to again. If critical (or *vital,*) as records are distinguished from the other records filed by company personnel, surely the result will be a smaller volume of records needing to be stored off-site. This, in turn, will reduce the costs of off-site storage.

Review of Records Management Standards

Of course, sorting through records and data with a Critical/Noncritical yard-stick may seem somewhat simplistic. For those who seek more refined categories, organizations for records managers and administrators and related fields have articulated several fairly sophisticated records classification schemes. Readers should obtain copies of records classification, retention and protection standards from the National Fire Protection Association in Quincy, Massachusetts and the Association of Records Managers and Administrators (ARMA) in Prairie Village, Kansas.

It is also a good policy to invite company auditors and legal staff to review record classification strategies. This will help to ensure that they comply with or incorporate special classifications required by law.

DEVELOP A BACKUP PLAN

Having developed an approved record classification strategy, the next step is to apply it to corporate information assets. The objective is to identify records and data that require protection both in user work areas and in the corporate data center or centers. In a small business, this may seem an insurmountable task; in a major corporation with worldwide offices, it can only be termed Herculean. The technique for accomplishing the task in either case is to obtain the direct involvement and assistance of work unit management. This may be accomplished by making work unit managers "owners" of the data and records they produce. Fortunately, this is exactly how the managers often perceive the situation.

Identifying Storage Requirements of Work Units

Some information on records and data requirements will have already been collected in the questionnaire completed by work unit managers at the outset of the project. Figure 7.2 provides a convenient form that may be distributed to work unit managers for their use in identifying and classifying records and for specifying known data retention requirements.

Looking at the form, the work unit managers are asked to indicate five items of information about each record they identify. First, the form uses a "Class" code that may be used for statistical reporting on data records retention. The work unit manager selects a number from Column A at the bottom of the form to specify the type of data or record that is being identified for retention. Then, the manager selects the appropriate numeric code from Column B to specify the time requirement for retention. The time requirement specified may be based upon practical assessments of record use or to fulfill some legal or audit requirement. This "Class" code—the combination of the alphabetical and numeric codes—is entered in the "Class" column of the form.

Next, the manager writes a brief description of the records and the type of "Media" on which the record is stored. Storage may be on paper, microform (film or fiche), electronic media (diskettes, tape reels, tape cartridges, tape cassettes or removable hard disk media), or optical disc.

RECORDS RETENTION WORKSHEET

INSTRUCTIONS: Complete this form to indicate the retention requirements for records produced or used by your work unit. Indicate the record class using the legend below. Indicate the media format of the record using the codes P for paper, E for electronic media, O for optical disc, or M for microform (fiche or film). Specify current volume in numbers of storage boxes for off-site storage planning (paper records only).

Class	Description	Media	Volume	Special Requirements

CLASS CODES: A class code consists of two entries, one from column A and one from column B below. (ex: 2B)

COLUMN A
1 -- Accounting/Fiscal Data
2 -- Personnel Records
3 -- Legal/Tax Records
4 -- Sales/Marketing Records
5 -- Manufacturing/Plant Records
6 -- Corporate Records

7 -- Purchasing/Procurement
8 -- Transportation/Shipping
9 -- Security
10 -- Medical
11 -- Other

COLUMN B
X -- 0-12 months revolving retention
A -- 1-5 year retention requirement
B -- 5-7 year retention requirement
C -- 7-10 year retention requirement
D -- 10-15 year retention requirement
E -- Permanent retention requirement

Figure 7.2 Records retention worksheet.

A column labeled "Volume" is for later use in assigning records to a specific storage crate or box which is stored off-site. The "Special Requirements" column specifies any special handling, preservation, or retention requirements not covered in the "Class" code definitions at the bottom of the page.

The worksheets provide the means to keep track of the records storage requirements of all company work units. They further provide a convenient list for use in conducting periodic reviews of work unit storage plans. A work unit manager can quickly review the list, remove items that no longer need to be retained, add new items, and so forth.

Identifying Storage Requirements of Data Centers

Of course, one work unit within the company with specialized backup requirements is data processing. To be effective, data backup in a data processing environment must occur on a daily basis; careful attention must be paid to data associated with critical systems and networks.

Figure 7.3 provides a basic form for tracking tape backup activity in data centers. (Tape is used generically here, since the actual backup media employed may be tape reels, tape cartridges, or some type of magnetic or optical removable media.)

Generally speaking, backups in data centers are of two types: full volume and incremental. One can make a complete copy of the system, including system software, utilities, applications software and data, creating what is often called a *full volume* backup. In many shops, the time requirements for accomplishing this task cause full volume backups to be taken weekly or even less frequently. The data center manager is thus asked to indicate at what intervals full volume backups are made.

Also of importance to off-site storage planning are answers to the questions: How many tapes (or other storage media) are used in the backup? How many copies of the full volume backup are produced? How frequently are backups tested for restorability?

The number of tapes produced in a full volume backup will partially determine the physical size requirements of the off-site storage facility. The number of copies that are made—more than one backup is typically made as a hedge against a non-restorable backup—will affect both the storage requirements and the frequency with which tapes must be rotated to the data center for periodic testing and verification of their restorability. Finally, the planned frequency of rotation will affect the tape handling and transport services that will be sought from an off-site storage strategy.

Obviously, in the time period between regular full volume backups, new data will be entered, applications will change, and in some cases even system software will be modified. These changes are typically backed up nightly to removable media in what is sometimes called an *incremental backup*. Two or more copies of the incremental backup are often made, with one or more copies stored off-site. Thus, another section of the form asks about the frequency of backups, the number of tapes involved, the number of copies made, and the frequency with which their restorability is verified.

TAPE BACKUP WORKSHEET

TAPE BACKUP SCHEDULES AND STORAGE REQUIREMENTS

FULL VOLUME BACKUPS: Full volume backups of the system are performed at regular intervals. The frequency of these backups is:

- ☐ Weekly
- ☐ Bi-weekly
- ☐ Monthly
- ☐ Bi-monthly
- ☐ Quarterly
- ☐ Other: _____

STORAGE REQUIREMENTS: Full volume backups currently require _____ tapes. The number of copies made of the full volume backup is _____.

VERIFICATION OF RESTORABILITY: Full volume backups are tested for restorability every _____ (months/weeks/days).

INCREMENTAL BACKUPS: Incremental backups of key system and data files are performed at regular intervals. The frequency of these backups is:

- ☐ Daily
- ☐ Every other day
- ☐ Weekly
- ☐ Bi-weekly
- ☐ Depends on Transaction
- ☐ Volume
- ☐ Other: _____

STORAGE REQUIREMENTS: Incremental backups currently require _____ tapes. The number of copies made of the incremental backup is _____.

VERIFICATION OF RESTORABILITY: Incremental backups are tested for restorability every _____ (months/weeks/days).

TAPE BACKUP METHOD AND SOFTWARE

BACKUP METHOD AND SOFTWARE: Incremental and Full Volume backups are created using (logical/physical image) methods. The software used to create and restore backups is:

Product Name:	_____
Version/Release:	_____
Manufacturer/Contact:	_____

Figure 7.3 Tape backup worksheet.

At the bottom of the form, the details of the backup method and backup software that are employed are recorded. The data processing manager should indicate whether logical or physical image methods are used to create backups by circling the appropriate selection.

With logical backups, backup software opens every dataset, reads each record individually, and processes the files in some logical order (i.e., by size of file). Logical processing is hardware-independent and will reorganize data during a restore in order to place data on working storage devices in the most efficient manner.

Physical image backup, on the other hand, operates at the hardware level and backs up data from disk drives in the fastest way possible. Backups and restores occur at as much as twice the speed of a logical backup, but there is no file reorganization in either the process of backing up or restoring files. Moreover, with physical image backups, restorations generally must be made to storage units that are of the same model as the original storage units—an important consideration when identifying a suitable mainframe recovery strategy.

Once the manager has identified the type of backup software that is used, it is important to record the name, release number, version number, and manufacturer of the software. More than one disaster recovery effort has been delayed because of differences between the software used to make backups and the software at the mainframe recovery facility that was to be used to restore backups. Ensure that contact information for the backup software vendor is recorded and that the vendor can be reached 24 hours a day for assistance should restoration problems arise.

Storage Requirements of Small Systems

Of course, between mainframe backup and work unit records storage, there may be a diverse array of other systems whose data also needs to be backed up. PCs, local area networks, and multiuser systems are the hallmark of modern decentralized computing. A number of software and hardware solutions exist for backing up these systems and networks.

In the world of decentralized systems, backups are most often made to diskette, but magnetic tape and removable hard disk are steadily growing in popularity. Partly, this is a response to the increasing size and decreasing price of hard disk drives: The larger the disk, the more cumbersome disk backup to floppy diskettes has become. Backing up a 20MB disk required 56 standard 360K floppy diskettes, or 14 high-density (1.2MB) floppies. Backing up an 80MB disk requires 24 boxes of 360K floppies or six boxes of high-density diskettes. Imagine the diskette requirements for the new multigigabyte drives that have become increasingly popular with LAN fileservers!

Another fact is: The disk operating systems of most popular personal computers provide an integral back up and restore facility that may be used to backup fixed disk contents to diskette. However, more and more users are turning to third-party software to do the job because of the slow speed and poor compression offered by the native backup/restore functions of DOS.

New backup software products offer four basic features that garner them the praises of small systems administrators and users.

- **Ease of Use**—Most backup and restore capabilities of native disk operating systems have a command-line interface requiring the user to specify backup parameters through the use of command-line *switches*. Third-party products, on the other hand, are typically designed with an extremely friendly interface that permits users to set backup parameters—including the specification of particular directories or files to be included or excluded, the setting of automatic file compression options, etc.—almost intuitively.

- **Automatic Function**—Many third-party packages also allow users to set up a schedule for automatic backup, so backups may be taken while the system is unattended or at off-peak hours. Some packages provide a memory-resident utility that will perform continuous backup "in the background" while users are working. Still, others allow the use of scripts (small user-defined programs in a simple macro language) to initiate backups of selected directories with only a few keystrokes. These features can reduce the time required to back up the small system.

- **File Compression and Security Encryption**—Many packages provide the means for compressing files as they are backed up so that more data will fit on the backup media. (Compression is accomplished via an algorithm that evaluates data, removes white space, and adds instructions for *uncompressing* the file via the same algorithm. Popular algorithms are artfully named "squeeze," "crunch," "smash," "shrink," etc.) Additionally, several products allow the user to encrypt the backup (or do so automatically), so a restoration may not be accomplished without a password or the original distribution diskettes.

- **Media Flexibility**—Another benefit of third-party software is its support for the wide range of backup devices available on the market. Many native backup programs will only output to diskette drives or to tape drives that use diskette controller boards within the computer (the so-called *floppy tape* systems). The data transfer rates to floppy devices are comparatively slow; a number of high-speed backup hardware subsystems are now available that cannot be used with native backup software. Third-party backup software is often provided with high-speed backup hardware to facilitate the use of its high-speed functions.

In light of these features, it is important to learn three things about the requirements of decentralized systems for purposes of off-site storage planning. For one, planners must identify how systems are currently being backed up—what hardware and software is used, how often are backups taken, and what is required for restoration. Additionally, the planner must learn how backups are currently being stored and whether any verification activity is performed to ensure restorability of the backup. Finally, the planner needs to discover what media is being used to store the backup and in what volume.

Figure 7.4 provides a worksheet for recording some of these items of information. It is also important to refer to the documentation collected during risk analysis to identify which systems are not regularly backed up that need to be and to identify any previously existing off-site storage programs that may be in effect.

BACKUP STRATEGY WORKSHEET
FOR SMALL SYSTEMS

LOCATION AND TYPE OF SYSTEM

WORK UNIT: _____

CONTACT: _____

TYPE OF SYSTEM: _____

DESCRIPTION _____

_____

_____

BACKUP STRATEGY

IS A REGULAR PROGRAM OF BACKUP CURRENTLY IN PLACE? YES/NO
IF YES, DESCRIBE THE BACKUP METHOD:

FREQUENCY OF BACKUPS:

FULL VOLUME? _____

INCREMENTAL? _____

NUMBER OF COPIES MADE: _____

DATA COMPRESSED? _____

DATA ENCRYPTED? _____

MEDIA USED: _____

QUANTITY OF MEDIA: _____

BACKUP HARDWARE AND SOFTWARE VENDOR CONTACT INFORMATION

BACKUP HARDWARE _____   BACKUP SOFTWARE _____

DESCRIPTION: _____   DESCRIPTION: _____

_____   _____

VENDOR: _____   VENDOR: _____

MODEL: _____   VERSION/RELEASE: _____

SERIAL NUMBER: _____   SERIAL NUMBER: _____

CONTACT: _____   PASSWORD: _____

   CONTACT: _____

_____

_____

Figure 7.4 Backup strategy worksheet for small systems.

Putting Together a Requirements List

Once data has been collected regarding the type and quantity of records to be stored off-site, the planner is ready to finalize storage requirements for use in soliciting bids from off-site storage vendors. This assumes, of course, that a commercial off-site storage strategy is the preferred strategy of the company. Commercial storage contracts and other strategies are explored in greater detail later.

For now, it is important to consolidate the information that has been collected to arrive at a clear idea of the storage resource that will be required for the company's critical data and records. Figure 7.5 provides a worksheet for this purpose.

Clearly, one requirement that must be defined is the aggregate volume of records and media that will need to be housed at the off-site storage facility. All worksheets must be totalled to obtain a working estimate of the volume of tapes, diskettes, discs, and boxes that will need to be stored on a permanent or long-term basis and on a shorter-term or rotating basis.

Special requirements for storage handling should also be identified. For tape media, this may include instructions for periodic testing and validation of data integrity and periodic tape rotation. For paper and microform records, special requirements may include environmental controls, periodic access for inspection, and occasional audits.

On the second page of the worksheet, a preliminary rotation schedule is worked out. The schedule may be simple or complex, but it will always contain information about where storage items are to be picked up, what storage items are involved, how often rotation must occur, and who will be involved in receiving deliveries or preparing returns.

Also, at the bottom of the page, authorized requestors must be designated for emergency delivery requests. This is done for two reasons. First, if an emergency occurs and the normal requestors are unavailable to authorize a delivery, someone else must be authorized to issue the request. The second reason for designating requestors is to prevent unauthorized individuals from gaining access to sensitive data and records that are stored off-site. Even in an emergency, appropriate levels of information security must be maintained.

OFF-SITE STORAGE REQUIREMENTS WORKSHEET

PERMANENT STORAGE

INSTRUCTIONS: Use Records Retention Worksheets and Tape Backup Worksheets to estimate storage capacity requirements for permanent off-site storage. Assume standard one cubic foot containers for paper and microform storage.

RECORD TYPE	VOLUME	UNIT OF MEASURE
TAPE REELS	_____	(# OF REELS)
TAPE CARTRIDGES	_____	(# OF CARTRIDGES)
PAPER FILES	_____	(IN CUBIC FEET)
MICROFORM	_____	(IN CUBIC FEET)
OPTICAL DISC	_____	(# OF DISCS)
OTHER MEDIA	_____	(IN CUBIC FEET)

SPECIAL HANDLING OF TAPES IN LONG TERM STORAGE:

Specify special handling requirements for tapes, including schedule for periodic validation of tape integrity, and periodic rotation of hanging tape.

SPECIAL HANDLING OF NON-MAGNETIC MEDIA:

Specify special handling requirements for storage other than tape, including environmental requirements, periodic access for inspection, and periodic rotation back to on-site work areas.

Figure 7.5 Off-site storage requirements worksheet.

TEMPORARY/ROTATING STORAGE

INSTRUCTIONS: Use Records Retention Worksheets and Tape Backup Worksheets to estimate storage capacity requirements for temporary or rotating off-site storage. Assume standard one cubic foot containers for paper and microform storage.

RECORD TYPE	VOLUME	UNIT OF MEASURE
TAPE REELS		(# OF REELS)
TAPE CARTRIDGES		(# OF CARTRIDGES)
PAPER FILES		(IN CUBIC FEET)
MICROFORM		(IN CUBIC FEET)
OPTICAL DISC		(# OF DISCS)
OTHER MEDIA		(IN CUBIC FEET)

PRELIMINARY ROTATION SCHEDULE

PICKUP FROM	DESCRIPTION	WHEN/HOW OFTEN	DELIVER TO

AUTHORIZED REQUESTORS FOR EMERGENCY DELIVERIES: Initial list, subject to change, of three individuals who may request emergency transport or access to stored records and data. Provide name, title, and driver's license or social security number:

Figure 7.5 Off-site storage requirements worksheet. (Continued)

OFF-SITE STORAGE STRATEGIES

Up to now, a subtle and unjustified assumption has been made that most readers will utilize a commercial vendor to accommodate their company storage requirements. In fact, commercial storage is one of several options that may be available to the planner. It happens to be the option most frequently employed because, for most companies, the conditions for secure, long-term storage may be most conveniently met by a commercial vendor. These conditions include:

- **Facilities**—For many companies, the cost of building and equipping with the appropriate environmental and access controls a special facility for storage is prohibitive. Not only should facilities offer humidity and temperature levels appropriate to the records and media they house, they should also feature fire protection, access controls, power backup, and other capabilities that ensure data and records are properly maintained. Moreover, the storage facility must be sufficiently distant from the company quarters where records and data are produced so as not to be affected by same regional disaster that impairs company operations.

- **Personnel**—The storage facility must also be staffed by knowledgeable personnel who have appropriate security clearances to be trusted with sensitive data and records. At many commercial facilities, personnel are subjected to rigorous background checks and are tested regularly for indications of drug or alcohol abuse that might point to a security risk. Few companies are positioned to hire a dedicated storage facility staff and to monitor their activities so carefully.

* **Transportation**—It is when data and records are moved to and from an off-site storage facility that they are most exposed to loss or damage. To transport records and media safely requires vehicles that are equipped with expensive environmental and security systems. Here again, few companies choose to purchase and equip several vehicles for media transport.

These are a sampling of the conditions that an off-site storage facility must meet to provide safe, secure quarters for critical records and data. As previously stated, commercial off-site storage vendors—which are plentiful around most major urban centers—are properly equipped and reasonably priced to provide the support that an average company requires.

In most companies, a contract is made with an off-site storage vendor. The contract specifies, among other things, how much permanent storage space (priced in cubic feet) is to be provided to the company and what stored records and data are to be rotated back to the company at stated intervals for review, validation, or replacement. The rotation schedule is typically timed to correspond to the backup schedule that has been implemented within the company, so that each delivery from the off-site storage company is met with a pickup of replacement records and data from the customer site.

Off-line versus On-line Storage

The description of off-site storage just provided is an example of an off-line strategy for backup and storage. That is, data is copied to a removable media, packed into containers or boxes, and shipped via car or van to a remote facility until it is needed for use or replaced with more current data. This is the strategy with which most companies are familiar.

The disadvantages of this strategy, however, are: (1) it can only be implemented cost-effectively if the vendor and the company are within about 50 miles of each other—thus, exposing both the original records and the copies to the same risk of regional disaster; and (2) the strategy entails the exposure of vital records and data to loss or damage by transporting them on highways in vulnerable automobiles and vans.

In response to these drawbacks of off-line strategies for storage, some companies have begun to apply the technologies of imaging and high-speed data communications to develop on-line strategies. In an on-line strategy, all critical data and records are transported electronically—via electronic data communications networks—to a secure storage location. This is the concept behind the *data vault*.

In data vaulting, a data center is connected to a remote data vault (basically, an off-site storage facility equipped to receive electronic data communications) via a high-speed digital network. (The data communications speed restrictions on analog networks often render them inappropriate for data vaulting.) In one scenario, backups, generated nightly at the company host, are moved to the remote vault during off-peak hours where they are dumped to tapes and stored in conventional tape storage facilities.

However, the premier implementation of data vaulting, as depicted in Figure 7.6, utilizes channel extension and a broadband digital communications facility to accomplish continuous backup. In operation, two writes are made by the company host of any data that is entered, modified,or deleted. One write is to local storage devices, the other is to a storage device (tape or disk) that is designated as installed to a particular channel on the mainframe. In actuality, the device on the channel is a digital communications unit called a *channel extender*. The extender forwards the data from the host to a remote channel extender at an off-site storage facility to which disk or tape devices are connected. The data is thus written to the remote device via the extended channel.

This technique is not only on-line but real-time as well. What this means is that an up-to-the-minute copy of all company transactions exists at the data vault (more often than not, the company's mainframe recovery center) at all times. Restoration of critical company systems may thus be accomplished speedily and completely in the event of an unplanned interruption.

While costly, the on-line benefits of data vaulting may be justifiable within the context of downtime costs. Also, with the increasing number of low-cost public network services that are being seen as both long distance carriers and local telephone companies are permitted to enter the business networking field, there is a good possibility that the most expensive element of data vaulting—digital facilities—will be dropping in price. Planners should keep themselves apprised of these developments.

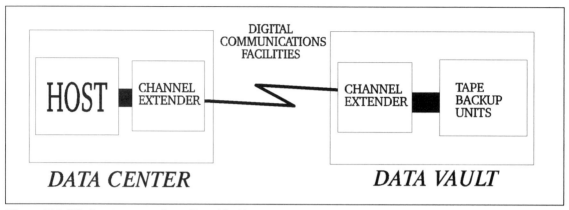

Figure 7.6 Off-site storage options: Data Vaulting.

Imaging Backup

Of course, on-line strategies are only appropriate for information storage in an electronic format. This has led many commentators to exclude paper and microform storage from discussions of the data vaulting solution—and incorrectly so.

To convert a document or microform into electronic form, it may be *imaged*—scanned and converted to a raster of black-and-white (or color) dots called *pixels*. This image raster may be stored electronically and recalled via a database application for review and printing. As electronic information, the document or microform image may be transmitted via electronic data communications and retained on-line at the data vault.

Many companies already use image systems and databases for functions previously performed using paper files or microform reader/printers. Legal tests have yet to be passed for the admissibility of imaged texts and documents in courtrooms or IRS hearings. However, complementing the permanent off-site storage of original legal documents with document images is an effective means for making these resources available to those who must access them to continue business in an emergency.

SELECTING AN OFF-SITE STORAGE VENDOR

Since the use of commercial off-site storage companies remains the prevailing method for businesses that develop off-site storage strategies, it is important to identify considerations that the disaster recovery planning team should keep in mind when evaluating the offerings of several vendors. Some of the basic considerations for selecting an off-site storage vendor were suggested in the section titled "Putting Together a Requirements List." Figure 7.7 provides an overview for further discussion.

As suggested in the figure, there are seven discrete areas to be evaluated within any commercial off-site storage vendor offering.

MEDIA

Can the vendor handle the volume and types of media that the company must store?

FACILITY

Do vendor facilities meet standards for design and protection articulated by ANSI, NFPA, ARMA, and/or ACRC?

TRANSPORTATION

Will media be transported to the vendor site in secure, unmarked, and environmentally-sound vehicles?

PERSONNEL

Are vendor personnel trained and experienced in proper media handling techniques? Are personnel subjected to periodic security reviews?

SECURITY

Are adequate physical access controls installed and is access restricted to vendor personnel?

ACCESS

Is 24-hour access provided? What is required to secure data from storage when needed?

LOGISTICAL SUPPORT

Does the vendor offer support in plan development or implementation?

Figure 7.7 Off-site storage vendors: Evaluative criteria.

1. **Media**—Media handling is receiving increased attention within the data processing community as the problem of failed restores from tape backups has grown to near-epidemic proportions. Unusable tapes are the major cause of disaster recovery plan test failures and a significant source of recovery delays in actual disaster situations. More than one company affected by the computer virus attacks that received widespread media coverage the late 1980s found that critical backups, required to restore virus-infected systems to their preinfection status, were corrupt.

    From the standpoint of media, off-site storage vendors should be assessed for their ability both to provide the required environment for long-term tape storage and to maintain media properly while it is stored. To assess storage facility environmental controls, planners can refer to guidelines published by the Association of Commercial Records Centers, the National Fire Protection Association, and the American National Standards Institute, and to storage guidelines from the media manufacturer itself. Vendors who comply with these guidelines often display certificates or otherwise indicate compliance.

    In the area of tape maintenance, planners should look for vendors who offer tape-cleaning services, periodic integrity tests, and regular tape reel rotation. However, as a backup, planners should ensure that long-term storage items are periodically returned to the data center and that test restores are conducted using the media.

2. **Facilities**—The formal standards previously cited for proper tape storage either imply or enumerate specific guidelines for storage facility design, as well. NFPA, ACRC, ANSI, and also the Association of Records Managers and Administrators, have articulated standards for facility construction. Such standards encompass fire ratings for structure walls, types of disaster alarm and suppression systems that should be employed, ambient temperature and humidity levels that should be maintained, and even locational considerations such as elevations, proximity to airports, etc. Planners should evaluate various storage facilities in light of both articulated standards and other construction codes enforced by state and local agencies.

3. **Transportation**—As previously stated, media and records are most vulnerable while they are being transferred from the company to an off-site storage facility. Most vendors offer their own transportation service. Ensure that transportation vehicles are equipped to maintain proper environmental conditions for the media that is being transported, that they offer fire suppression systems, and that the vehicles are unmarked to conceal their contents.

4. **Personnel**—Personnel who transport, handle, and maintain storage require training in media handling and should be subjected to background checks and on-going security reviews. Vendors should be able to supply proof that they have taken measures to ensure that stored records and media will not be disclosed, sold, or otherwise released by vendor personnel to competitors or unauthorized persons and that vendor staff possess

the knowledge and skills to prevent damage to stored records and media due to mishandling.

5. **Security**—Physical access controls (and electronic communications security measures, in the case of electronic data vaults) should be scrutinized by the disaster recovery planning team. The vendor should demonstrate that media and records are secure both when in storage and during transport. Planners should watch out for vendors that allow unsupervised access to storage areas by outside persons or customers. Moreover, vendors should be bonded or otherwise contractually bound to compensate customers for the loss, damage, or disclosure of records and data in their charge.

6. **Access**—Just as access to records and data should be restricted only to authorized individuals, accompanied by vendor supervisory staff, so too must access be sufficiently flexible to permit access to records and media in an emergency. Vendors should permit the designation of several authorized requestors who may request delivery of (or obtain in person) records and media required in a recovery situation. That access must be forthcoming on a 24-hour basis. Therefore banks, savings-and-loans, and other depositories with normal business hours are often excluded from consideration as off-site storage vendors.

7. **Logistical Support**—More and more off-site storage vendors are beginning to offer services intended to customize storage services to the customer's disaster recovery plan requirements. For example, some vendors offer to develop and maintain the portion of the disaster recovery plan related to off-site storage programs, using their own computer systems to track rotation schedules and to maintain approved requestor lists. Others will facilitate the transport of media to a remote mainframe or user backup facility, maintaining storage crates on-site that will be quickly loaded and delivered by the vendor to the local airport for prompt shipment. These services generally carry an additional cost to the normal spectrum of off-site storage services and planners should evaluate them carefully to ensure that appropriate services are purchased to support the disaster recovery requirements of the company.

Types of Off-Site Storage Vendors

There are numerous types of off-site storage facilities. Moving and storage companies, commercial storage facilities, financial institutions, specialized data and records storage companies, and even bomb-proof underground vaults have all been utilized as storage locations. Keeping in mind the requirements previously listed—particularly media requirements, facility design requirements, security requirements, and access requirements—will aid planners in separating acceptable from unacceptable facilities.

Generally, businesses in most medium-to-large cities have a variety of off-site storage vendors from which to choose. Companies located in smaller cities that lack commercial, media-rated facilities, may be able to make use of cooperative agreements with other companies.

Sharing storage with a neighbor is a kind of "shell game" approach to the possibility of a disaster. The assumption is that records and data stored off-site will not be subject to a local disaster such as a fire that consumes the originating company. However, regional disasters can make local data and records and off-site records and data prone to loss. This is the same whether a commercial off-site storage arrangement or a cooperative approach is employed.

Locating a Vendor

Off-site storage companies are listed in the telephone directory and often advertise in local newspaper business sections. Referrals from business associates or local disaster recovery professional associations are another source of information about storage vendors in the planner's vicinity.

SOLICITING BIDS AND EVALUATING OPTIONS

As with all products and services that are purchased from third parties, a formal process of specification and bid evaluation is a good policy for vendor selection. Following the definition of storage requirements and the identification of several potentially suitable vendors, the disaster recovery planning team must solicit bids from vendors.

Vendors should respond with a proposal of services, a contract, and the names of several reference accounts that can be contacted for additional information about the vendor's performance. The vendor should also provide:

1. Certifications of facility compliance with standards for media storage and appropriate building codes

2. Certifications of bonding and/or insurance

3. Descriptions of employee training and personnel security measures

4. A tentative schedule for media transport and maintenance, with all costs clearly indicated

Once documentation has been received, the planning team should follow up on certifications with their issuing authorities. It is also a good idea to consult the Better Business Bureau, governmental consumer protection agencies, state licensing organizations, and others that can provide an objective history of the organization from the standpoint of past business practices.

Another obvious step is to consult with client references provided by the vendor. Ask about contract performance, the client company's storage volumes, and any vendor mishaps and how they were handled. Planners should strive to obtain as complete a picture as possible of what the client's requirements are (for comparison to their own) and how the vendor has met those requirements.

Finally, the planning team should schedule visits or tours of each vendor facility. Tours should include transport vehicle inspection, inspections of storage facilities and maintenance equipment, and discussions with managers and staff.

DOCUMENTING VENDOR BIDS

For audit purposes, it is beneficial to document the planning team's decisions about vendors. Each team member involved in vendor selection should write a brief text summarizing his or her observations about the competing vendors and recommending a specific vendor for adoption.

Ultimately, the decision about which vendor to use may not be based strictly on price. The quality of vendor facilities, services, personnel, etc. may argue in favor of a more expensive vendor on the grounds of perceived reliability.

The project leader may be required to render the final decision on off-site storage vendors, taking into account the recommendations of team members, his or her own observations, and the budgetary constraints on the planning project. The project leader should also attempt to negotiate all prices on contracted services, since some vendors have substantial latitude in their contract pricing depending on storage volumes, projected growth, and other factors.

Finally, some provision should be made for both the initiation of off-site storage and the regular auditing of vendor performance. Initiation steps include conducting meetings with user departments to advise them of routine schedules for pick-up and delivery and emergency procedures, and to assign compliance responsibility to specific persons in each work unit. The first few regular pick-ups should be monitored to ensure that responsible parties are participating and that everyone understands what is required when.

Plans should also be formulated to test the off-site storage program on an on-going basis. At regular intervals, the vendor contract should be audited to ensure that schedules are being observed, that appropriate vehicles are being used, and that facilities have not been downgraded. With or without prior vendor notification, the off-site storage plan should be tested. Ensure that tests of the disaster recovery plan include the acquisition of backups, supplies, and other materials stored off-site from the vendor. Simulate an actual emergency and observe how the off-site vendor performs its emergency delivery services.

SUMMARY

There are several variations on the depiction of off-site storage backup presented in this chapter. In some cases, companies may elect to store permanent backups at their user or mainframe backup site. One vendor of mainframe backup services recommends that operating system backups that have been constructed in the course of a formal test of the disaster recovery plan be retained at the backup facility and updated only when another test occurs. Assuming that testing is conducted on a fairly frequent basis, or whenever a change is made to system software, this strategy has considerable merit.

Another variation involves the off-site storage requirements of work units when the work units are distant from each other. It is quite possible that multiple vendors may need to be used and that work unit managers will need to evaluate and select their own vendors, as well as perform their own requirements analysis. When this is the case, the planning team may need to prepare advisory documents to assist work unit managers in evaluating and selecting the best possible vendor for the job. Then, they will need to obtain copies of contracts and storage lists from each work unit and ensure that work unit managers are regularly auditing their storage strategies.

Off-site storage and the installation of disaster prevention capabilities are keys to reducing the likelihood of a non-recoverable disaster. If the planning team accomplishes nothing else but these two phases of disaster recovery planning, it will have made a substantial contribution to the integrity and survivability of corporate business functions.

Effective backup and off-site storage programs are critical to the timely restoration of business functions in the wake of a disaster. These programs are closely coupled with individual strategies for system, network, and end-user recovery and should be reviewed in the context of recovery requirements associated with these strategies once they are formalized.

Checklist for Chapter

1. Develop criteria based on legal requirements, tax law requirements, records management guidelines, corporate records retention policies, and insurance requirements for assigning criticality to documents.

2. Complete records retention worksheets for end-user departments.

3. Identify backup procedures and schedules for distributed and centralized systems. Determine adequacy and completeness of procedures.

4. Prepare an off-site storage requirements list.

5. Identify an off-site storage strategy. Solicit vendor bids. Document the preferred strategy.

Systems Recovery Planning

SETTING OBJECTIVES FOR SYSTEMS RECOVERY

Systems recovery—the recovery of computer platforms, their operating system and application software components, and their local peripheral device network—is among the most well-documented aspects of disaster recovery planning. For the past 20 years, much of what was termed disaster recovery planning was synonymous with planning for the restoration of the corporate mainframe. This has resulted in an established set of mainframe recovery alternatives supported by a mature industry of vendors offering an array of system recovery facilities.

However, just as mainframe recovery strategies have evolved into pat text-book solutions, the world of computing has changed. Businesses are increasing the power and complexity of the mainframes they use and, at the same time, decentralizing automation to place greater resources directly in the hands of those who use them. The disaster recovery ramifications of these changes are numerous. They include:

- **The Multiplication of Recovery Targets**—The number of separate systems that must be recovered in order to recover business functions is increasing because of the proliferation of minicomputers, local area networks, and microcomputers that handle critical information functions.
- **The Diversification of Operating Systems and Platforms**—Decentralized automation has created a new field of computer science—systems integration—that seeks strategies for combining disparate and otherwise incompatible systems to facilitate their interoperation. Disaster recovery planning must borrow from systems integration to develop strategies for porting critical applications to fewer platforms in order to expedite recovery.

- **The Absence of a Central Standard on Technology Acquisition or Application Development**—In many companies, decentralized automation has led to the abandonment of company-wide technology acquisition and application development standards. This complicates the efforts of the disaster recovery planning project to integrate applications on fewer platforms in the event of an emergency. In more and more companies, the need for hardware and software standards that will facilitate integration is being championed by disaster recovery planners, auditors, and others who are concerned about the ability of the company to recover an increasingly sophisticated and disparate set of computer systems.

These are the milieu issues that the disaster recovery planning team needs to address as it develops strategies for systems recovery. However, these efforts, while they may bear useful results in the long term, cannot be allowed to displace immediate efforts to identify critical systems and to develop strategies for recovering them following a disaster.

Developing Target Systems Recovery Inventories

The determination of which systems are critical flows from the risk analysis conducted earlier in the project life cycle. In the risk analysis phase, business functions were scrutinized to reveal their relative importance to the organization. Basically, business function criticality was defined as the cost to the company resulting from an interruption of a business function for a protracted period. Critical systems were those systems that supported critical business functions either directly or indirectly.

Having identified which systems (or rather which application/hardware platform combinations) support critical business functions, the project team needs to inventory the hardware and software components that their recovery strategies will seek to restore. Figures 8.1 and 8.2 provide a format for performing the inventory as it pertains to software and hardware in the company data center.

In Figure 8.2, an extensive listing is made of data-center software. Working in conjunction with data-center management, planners must identify the operating system software and other utility packages—including security, job management, communications, and so forth—that comprise the software nucleus of the mainframe. The remainder of the form is used to record individual off-the-shelf and custom-developed applications software that serve as resources for critical business functions.

Hardware components of the data-center platform are covered in Figure 8.1. This form and attached configuration diagrams or other system documentation, describe the CPU, its backplane and storage configurations, attached peripherals, communications equipment, and other hardware.

These provide a baseline for the configuration that will need to be recovered in the event of a disaster. However, the baseline is not complete until the recovery requirements of other decentralized systems are examined and integrated into a minimum equipment configuration (see later section in this chapter).

DATA CENTER HARDWARE INVENTORY

CPU DESCRIPTION: *(Indicate make/model, memory size, special features, etc.)*

BACKPLANE CONFIGURATION: *(Identify channel configurations, connections)*

Page 1 of 3

Figure 8.1 Data center hardware inventory.

STORAGE CONFIGURATION: *(Indicate DASD make/model, capacities, etc.)*

_____
_____
_____
_____
_____
_____
_____
_____
_____
_____

PERIPHERALS: *(Indicate makes/models, mode of connection, special features, etc.)*

_____
_____
_____
_____
_____
_____
_____
_____
_____
_____
_____
_____
_____
_____
_____
_____
_____

Figure 8.1 Data center hardware inventory. (Continued)

COMMUNICATIONS FACILITIES: *(Indicate make/models for modems, etc.)*

_____
_____
_____
_____
_____
_____
_____
_____
_____
_____
_____

OTHER EQUIPMENT: *(List other data center hardware by type and function)*

_____
_____
_____
_____
_____
_____
_____
_____
_____
_____
_____

ATTACHMENTS: *(Attach configuration diagrams and other documentation)*

Please return this form and all attachments to:

Jon William Toigo
Disaster Recovery Coordinator
MIS Division
Room 123

Figure 8.1 Data center hardware inventory. (Continued)

DATA CENTER SOFTWARE INVENTORY

SYSTEM SOFTWARE

OPERATING SYSTEM: _____
LICENSE/SERIAL #: _____
VERSION/RELEASE #: _____
MANUFACTURER: _____
CONTACT: _____

UTILITIES

SECURITY

APPLICATION: _____
LICENSE/SERIAL #: _____
VERSION/RELEASE #: _____
MANUFACTURER: _____
CONTACT: _____

JOB MANAGEMENT

APPLICATION: _____
LICENSE/SERIAL #: _____
VERSION/RELEASE #: _____
MANUFACTURER: _____
CONTACT: _____

COMMUNICATIONS

APPLICATION: _____
LICENSE/SERIAL #: _____
VERSION/RELEASE #: _____
MANUFACTURER: _____
CONTACT: _____

Attach additional pages as needed

Figure 8.2 Data center software inventory.

OFF-THE-SHELF APPLICATIONS

APPLICATION: _____
LICENSE/SERIAL #: _____
VERSION/RELEASE #: _____
MANUFACTURER: _____
CONTACT: _____

APPLICATION: _____
LICENSE/SERIAL #: _____
VERSION/RELEASE #: _____
MANUFACTURER: _____
CONTACT: _____

APPLICATION: _____
LICENSE/SERIAL #: _____
VERSION/RELEASE #: _____
MANUFACTURER: _____
CONTACT: _____

APPLICATION: _____
LICENSE/SERIAL #: _____
VERSION/RELEASE #: _____
MANUFACTURER: _____
CONTACT: _____

APPLICATION: _____
LICENSE/SERIAL #: _____
VERSION/RELEASE #: _____
MANUFACTURER: _____
CONTACT: _____

Attach additional pages as needed

Figure 8.2 Data center software inventory. (Continued)

CUSTOM-DEVELOPED APPLICATIONS

APPLICATION: _____
PURPOSE: _____
LAST UPDATED: _____
CONTACT: _____

APPLICATION: _____
PURPOSE: _____
LAST UPDATED: _____
CONTACT: _____

APPLICATION: _____
PURPOSE: _____
LAST UPDATED: _____
CONTACT: _____

APPLICATION: _____
PURPOSE: _____
LAST UPDATED: _____
CONTACT: _____

APPLICATION: _____
PURPOSE: _____
LAST UPDATED: _____
CONTACT: _____

APPLICATION: _____
PURPOSE: _____
LAST UPDATED: _____
CONTACT: _____

Attach additional pages as needed

Page 3 of 3

Figure 8.2 Data center software inventory. (Continued)

Time Requirements

Obviously, the primary objective of system recovery is the restoration of applications that support critical business functions. However, as previously suggested, there are two corollaries to this objective: A satisfactory recovery strategy must be able to (1) accomplish restoration quickly and (2) do so at a manageable cost.

The first factor, time, is extremely important in developing a system recovery strategy. During risk analysis, data was collected regarding the costs to the company of outages of various durations. In most companies, it is time—more specifically, the length of time that a company cannot perform critical business functions—that distinguishes an inconvenient interruption from a disastrous one.

Based on risk analysis, it can usually be demonstrated that the costs of downtime and restoration increase almost exponentially as the length of downtime increases. In risk analysis, this relationship is expressed in terms of a recovery threshold, a so-called "maximum allowable downtime." Maximum allowable downtime is simply the maximum amount of time that the company can be deprived of critical information processing resources and still successfully recover.

Maximum allowable downtime is usually expressed in hours; the industry rule-of-thumb is 48 hours. However, some companies find that they can withstand a more prolonged outage, while others discover that their vulnerability to even short-term interruptions is much greater than the industry norm. Whatever the maximum allowable downtime threshold identified by the project team's risk analysis, it places a necessary condition upon recovery strategies. All recovery strategies—including systems recovery—must be able to be implemented within the stated timeframe.

Cost Requirements

The second requirement of the system recovery strategy is that it must have a manageable cost. A variety of options exist for mainframe recovery (see later chapter sections)—from the extremely low-cost, laissez-faire approach (in which the company waits for a disaster to strike, then finds a backup capability with which to recover) to the extremely expensive full redundancy option (in which a redundant data center is constructed and maintained in full readiness against the possibility of an interruption). In between are a range of alternatives whose reliability and pricing run from relatively inexpensive to quite costly.

Practically speaking, the cost of the systems backup strategy should not exceed the estimated loss from a disaster. However, this annual loss potential is virtually impossible to demonstrate to management. Thus, planners will need to assess all options and present the least costly option to management, armed only with a justification based upon the confidence that the strategy provides and a clear presentation of what the company could lose if disaster struck the company unprepared. From this perspective, it is easy to see why a moderately priced commercial mainframe backup facility is becoming the strategy of choice for many companies concerned about disaster recovery planning.

Systems Backup Strategy Development

Having examined the basic objectives and constraints on systems backup strategies, an overview of the activities that comprise this phase of the disaster recovery planning project is needed, as shown in Figure 8.3.

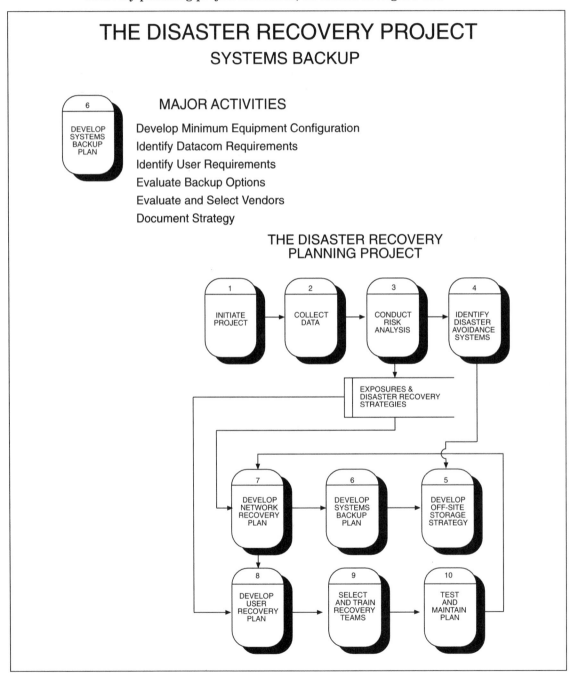

Figure 8.3 Systems backup phase.

- **Develop Minimum Equipment Configuration**—Using data collected as part of risk analysis, it is necessary to define the most streamlined configuration combining all of the critical applications that the systems backup strategy seeks to recover.
- **Identify Data Communications Requirements**—Because application processing in the backup system will be conducted at a remote location (that is, at some location other than the normal host site), it is necessary to develop an understanding of how this will impact normal communications with the system. Requirements for remote user terminal and peripheral networks will need to be identified as well as other applications-related communications needs.
- **Identify User Requirements**—Depending on the nature of the disaster, users may remain at their normal work locations or they may be moved to another location to perform work. A formal user recovery strategy is articulated later in project development. However, data needs to be collected at this stage to define what resources will be required at the user recovery location to reestablish processing channels and to continue work.
- **Evaluate Backup Options**—Having identified the configuration and other requirements for the backup system, next the planning team must look for a mainframe backup strategy that will fulfill those requirements, that may be implemented within the maximum acceptable downtime threshold, and that will not cost more to maintain than corporate management is willing to spend.
- **Evaluate and Select Vendors**—The fact is that more and more companies are selecting commercial mainframe recovery vendors as their systems backup solution. In recognition of this fact, planners need to know what to look for in a vendor and how to select the vendor that best meets their needs.
- **Document Strategy**—The output of the systems backup strategy is a set of procedures describing the strategy. These procedures document how the strategy will be implemented and become an integral component of the paper disaster recovery plan.

DEVELOP A MINIMUM EQUIPMENT CONFIGURATION

The decentralization of corporate computing has led to an increased concern about the potential for interoperability between disparate systems. In the mid-1980s, when decentralization first came to be recognized as an alternative to centralized computing, it was driven by several factors. Not the least of these was the stated need within company work units for more computing power and applications software.

Many centralized data processing or information systems departments found themselves unable to address this need quickly enough to satisfy work unit managers. The trade press and the small systems industry pointed to growing application backlogs and cost inefficiencies in centralized computing. This, combined with the increasing availability of turn-key minicomputer and low-cost microcomputer systems, led to the proliferation of smaller systems within the company.

The acquisition of these systems and the development of applications to run on them proceeded, often without regard for common standards or future integration requirements. The end product was a mix of operating systems and platforms that resisted integration attempts. The resultant problems were manifest: poorly documented or undocumented custom software, user requirements outstripping the capabilities of their small systems, hardware investments exposed to unpredictable upgrade paths and technological dead-ends, and so forth.

By the end of the 1980s, an entire industry had developed around *systems integration* and *enterprise processing*, seeking to solve operational inefficiencies wrought by the early years of decentralization. Where smaller systems have retained their usefulness, products have been introduced to support their continued operation but also to facilitate their integration into more capable platforms.

The point is that disaster recovery planners may confront multiple, disparate, nonintegrated systems as they set about developing recovery strategies. However, the same technologies that are available to support the integration and portability of applications for production environments may also be harnessed to develop a consolidated disaster recovery system. This, in a nutshell, is the definition of minimum equipment configuration.

The concept of a minimum equipment configuration may be implemented in a variety of different ways depending on the requirements and preferences of the disaster recovery team. In its simplest manifestation, an alternative configuration may consolidate multiple applications on a single platform. For example, a LAN or multiuser system may be used to host a number of applications normally operated on standalone PC systems. Ideally, however, the use of *portable coding* in custom software applications designed for smaller systems will facilitate their integration with and operation from a single mainframe platform in disaster recovery mode.

Initially, planners may need to develop a strategy that encompasses the recovery of several systems. Many mainframe recovery vendors will, for a price, make available at the recovery facility whatever small systems equipment the customer may require. However, wherever opportunities exist for platform consolidation, planners should seize them as a means for reducing the costs and logistical complications of the recovery strategy.

Ultimately, planners should encourage and endorse efforts at establishing corporate-wide standards on technology acquisition and software development that will allow applications to be moved from several small systems onto a single larger platform for recovery purposes. Moreover, the disaster recovery planning team should make available to company systems professionals all information on systems and networks obtained in their risk analysis and subsequent data collection activity. Chances are good that a comprehensive database, defining all installed systems, their purposes, and their applications, does not exist anywhere in the company given the fact that few projects have disaster recovery's company-wide focus.

Figure 8.4 provides a sample minimum configuration worksheet. In this case, the worksheet approaches consolidation from a simplistic point of view.

MINIMUM EQUIPMENT CONFIGURATION WORKSHEET
(Microcomputer Application Consolidation)

INSTRUCTIONS: Use this form to identify the requirements of critical microcomputer-based application software. Requirements will be summarized to define an alternative platform to consolidate recovery hardware requirements.

APPLICATION NAME	WORK UNIT	CPU/RAM	DISK REQS	SPECIAL REQS
	TOTAL RAM		TOTAL DISK	

Figure 8.4 Minimum equipment configuration worksheet.

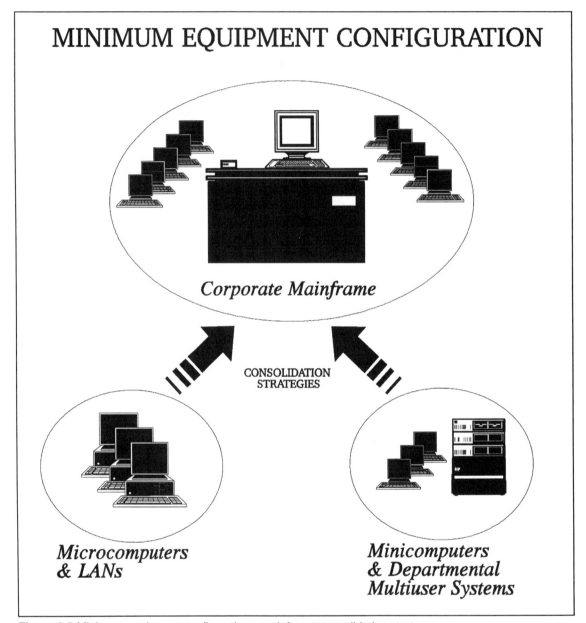

MINIMUM EQUIPMENT CONFIGURATION

Corporate Mainframe

CONSOLIDATION
STRATEGIES

*Microcomputers
& LANs*

*Minicomputers
& Departmental
Multiuser Systems*

Figure 8.5 Minimum equipment configuration—mainframe consolidation strategy.

The planner can use this form, together with terminal usage profiles and business function survey data, to list microcomputer-based applications in various work units. Both the applications and the work units employing them are listed on the form. Then, for each application, the CPU and memory requirements of the application, as well as the disk storage requirements for both application program and data files, are entered. In the last column, special requirements of the application (special video displays, input devices, peripherals, etc.) are recorded. Based on this information, as well as indicated usage of the application, a new hardware configuration—combining several platforms into one—may be developed.

A minimum equipment configuration in this case may consist of a multiuser system—an Intel 80486 or Pentium microcomputer running a multiuser operating system allowing for multiple concurrent DOS sessions—or a LAN—utilizing a fileserver and inexpensive diskless workstations. Depending upon the frequency of usage, it may also be possible to install a single, well-equipped micro that several users can share.

The objective, as illustrated in this simple example, is to consolidate applications onto fewer computer platforms. In so doing, critical automation resources may be able to be provided less expensively and with fewer logistical complications.

Once all critical applications (including those that reside on micro, mini, and mainframe platforms) have been evaluated and opportunities for minimum equipment configuration have been explored, summaries of the consolidation strategies should be prepared. These summaries provide the details of configurations that will be installed either at the user recovery location or at the mainframe backup site. The former comprises a user recovery component of the systems recovery strategy; the latter, illustrated in Figure 8.5, will be the basis for selecting the mainframe backup option best suited to company requirements.

Review Subject Systems

With the concept of minimum equipment configuration defined, it should be noted that formulating such a systems configuration can be a technical challenge. Expertise both from within the company and from vendors may be required to develop the most recoverable mix of standalone and combined system platforms.

This process begins with a review of all subject systems identified in the risk analysis phase. The planner needs to inventory mainframe, midrange, and microcomputer systems, determine whether the work they perform is critical, and determine parameters on their usage that may allow multiple systems to be consolidated into a fewer number.

Mainframe-based

Despite pundit predictions about the demise of the corporate mainframe, big iron continues to find use in corporate data processing. Mainframe-based systems traditionally provide a somewhat simplified target for recovery because of their centralized nature. Most current disaster recovery planning focuses on the mainframe recovery process and an established industry exists to support a range of mainframe backup strategies. To avail the company of these services, first, mainframe systems need to be inventoried and critical hardware components identified.

Also, it is increasingly the case that mainframes are part of larger distributed processing schemes. For example, mainframes often participate in corporate local area networks (LANs) where they provide centralized database or common storage functions. Channel extension is frequently used to extend peripherals away from the backplane of the mainframe and off the raised floor.

These connectivity configurations and their purposes must be clearly understood before consolidated mainframe configurations can be discerned.

Minicomputer-based

It is a fact that data stored on midrange or minicomputer platforms is increasing at a far greater rate than is data stored on mainframe-managed storage. Some vendors have begun to provide specialized recovery strategies, patterned on mainframe recovery methods, for companies that use these systems. As with the mainframe, hardware specifications and backplane configurations must be known before consolidation efforts can be made.

LAN-based

Workstations, microcomputers, and even minicomputers and mainframes are increasingly connected in local area networks. Most corporate LANs provide baseline services such as e-mail and file transfer, but more and more client/server and workgroup applications are being supported over LANs. If an analysis of the LAN reveals that it is used primarily for convenience, rather than database access or other forms of distributed processing, it may be possible to dispense with the LAN altogether in the disaster recovery system design.

Microcomputer-based

Microcomputers are at the same time commodity and special-order items. Some users require heavily configured micros to perform mission-critical work, while others use their PCs to run a word processor or spreadsheet, or to obtain access to a mainframe legacy application. PC configurations will need to be examined to determine what work they support and whether fewer machines with simpler configurations will meet recovery needs adequately.

Identify Critical Applications

With platforms inventoried, the next step in defining an overall recovery configuration is to review critical applications. Criticality was determined as part of the risk analysis. Planners need to revisit the analysis and associate applications with platforms to identify critical support requirements for each application.

This analysis will likely reveal that the systems and networks installed at the company have different restoration priorities. The planner needs to determine which applications must be available as soon as possible following an interruption and which can be delayed. This, in turn, will determine what system capabilities need to be present immediately and which can be provided later in the recovery period.

Identify Required Peripherals

Systems comprise not only host processors, but also peripheral devices for data input and output. Some devices, like the hosts themselves, are critical and must be available for immediate use in a recovery situation. Others may be required only periodically or their functions may be able to be consolidated.

Explore Opportunities for Integration/Consolidation

With hardware and application priorities established, the next step is to meet with company and vendor technical specialists to identify opportunities for integrating or consolidating resources on fewer platforms. In this context, integration refers to using a single host to operate several virtual systems. Consolidation refers to the combination of multiple applications to run in a single host environment.

Integration is possible on a large processor having the capability to support several partitions or operating environments. In many larger mainframes, it is possible to provide a UNIX environment in one partition, a PC DOS environment in another, a midrange operating system in another, and several mainframe operating environments in others. In this way, the applications that normally operate on a variety of systems can be integrated onto a single hardware host. They become virtual machines operating within a single physical unit. This might provide a means to minimize the numbers of workstations or PCs that need to be provided for recovery in the short term.

Consolidation occurs when multiple applications are ported to a single host, where their functions can be provided acceptably for a period of time. For example, it may be possible to consolidate the applications on multiple smaller systems—such as those normally connected in a LAN—by establishing a multi-user system with a single, highly configured workstation and several diskless workstations.

Of course, integration and consolidation options need to be evaluated on the basis of their ease of implementation and economy of operation. However, planners must also consider the ease and familiarity of use of these alternatives. New configurations, however economical, will not contribute to recoverability if they require significant adaptation and learning by end users. In general, the more that a recovery system looks and behaves to the end user like the familiar operations system it replaces, the easier it will be to perform required work.

Determine Usage Parameters

Usage is where the minimum equipment configuration option is put to the test. In addition to providing a physical and/or virtual replacement for familiar production systems, the recovery system needs to take into account other factors relating to its use. These factors include the numbers of users who must have access to the systems, access requirements in terms of terminals and communications, and the availability requirements for the systems.

Skeleton User Population

Most recovery scenarios do not anticipate system access by the entire complement of end users accessing production systems during normal operations—at least not in the initial stages of the recovery process. Using the terminal usage profile, business recovery requirements, and business function survey data, it is possible to determine the number of personnel that will serve as a *skeleton crew* to provide essential operations support using the recovery system.

The skeleton crew may be described in several ways. Planners may wish to identify numbers of users by application, by platform, or by work area. Ultimately, for planning, this number will need to be refined so that access resources and schedules can be delineated.

Minimum Terminal Network

With the user population associated with the recovery system defined, it is necessary to relate this information to access device requirements. In other words, planners need to ensure that the number of terminals or workstations are provided that users require to perform productive work.

In many scenarios, dumb terminals or simple PCs with terminal emulation software are preferred because of their widespread availability. Highly configured workstations are more difficult or expensive to provide in the timeframes required for recovery.

Data Communications Requirements

New configurations may also entail new communications requirements. In fact, unless end users are being recovered at the same location as the recovery systems, some provision will need to be made for remote dial-up or dedicated access. In some planning projects, the requirements for data communications access are recorded at this point for later consideration under network recovery planning. It may be preferable in other cases to document all data communications hardware and line requirements during the systems recovery planning phase so they can be included in the requirements definition submitted to vendors of system backup products and services.

Shift Scheduling

Determining skeleton crew number, terminal/access device requirements, and data communications access requirements is only part of the task of describing recovery system use. Planners also need to consider the practicality of scheduling system use as shift work.

In normal operations, users may require access to applications during a single *shift*—their typical eight-hour day. Subsequent system operations may consist of processing daily batches and performing system administration functions such as backups. In an emergency mode of operations, users may be scheduled in shifts to maximize the system use while minimizing access resource requirements. Shift planning may help planners to determine the optimal minimum equipment configuration—one that enables critical work to be performed while guaranteeing that adequate windows are preserved for batch and administrative processing.

DOCUMENT A MINIMUM EQUIPMENT CONFIGURATION

Once the analysis described in the previous section has been performed, the minimum equipment configuration decided upon needs to be documented. Doing so provides a record for auditing and testing as well as a "shopping list" to facilitate negotiations with vendors of disaster recovery services.

Hardware Requirements

A hardware requirements description, similar to the Data Center Hardware Inventory, should be developed for the recovery system. It should include all processors, storage devices, tape devices, controllers, and other equipment required for the configuration. The more specific this description is, the easier it will be to equip a backup facility in advance of a disaster or to locate a vendor to supply the required configuration following an interruption.

Software Requirements

Using a form similar to the Data Center Software Inventory, list all off-the-shelf and custom-developed software products that are required for installation on the backup system or systems. You may wish to note whether the software files are part of the data included in routine system backups or whether they will be contained on separate media. It is also important to indicate whether a separate version of the software has been prepared for use with the recovery system.

In addition to listing the software, it will be important to cross-reference the software to system resources to determine whether all platform requirements for the software application have been met in the recovery system configuration. Once recovery software has been defined, it will be essential to install the software as part of testing to ensure that no configuration requirements have been overlooked.

Data Communications Requirements

List all hardware requirements for connecting input and output devices to the recovery systems. Terminal emulation devices, DSU/CSUs, channel extenders, modems, switching equipment, routers, bridges, gateways, and other devices should be included.

Calculate the bandwidth requirements for the anticipated access via data communications. This may not be possible until a user recovery strategy has been formalized. Bandwidth requirements will be provided to the network recovery strategy planning process to determine what types of lines or facilities are required to recover connections between the systems backup location and other corporate work units.

User Requirements

User requirements need to be articulated in two parts. First, a complete accounting should be made of input and output devices that will be used by system users. These may be delineated by work unit, application or associated host system. The method used to connect devices to the host should also be noted.

Together with the device accounting, planners should articulate a systems usage schedule. The schedule should indicate when application use is anticipated, by whom, and on what basis (by shift or other basis). This schedule should be sufficiently detailed that system administrators can use it to develop a rudimentary run book identifying how resources will be allocated to on-line processing, batch processing, and administrative functions within a processing day.

Assumptions

Finally, planners should record all assumptions underpinning the minimum equipment configuration that has been defined. This section should include assumptions about the host operating environment, predictions about system performance, and any expectations or requirements drawn from or assigned to documented network and user recovery strategies.

EXAMINE RECOVERY OPTIONS

Once a minimum equipment configuration has been defined and documented, it is necessary to define a strategy for implementing the configuration at an alternate site should the production system and/or site become impossible to operate or occupy. Fortunately, considerable case study data is available to support an analysis of system recovery options. This section summarizes the major options and examines the costs and benefits accrued to each option.

Options (Ordered by Confidence Level)

System recovery strategies are well-established and tested. These strategies may be arranged on a confidence spectrum that indicates the performance of the strategy when implemented.

Redundant Facilities (High Confidence)

The strategy for system backup that affords the highest measure of confidence for a complete recovery of critical system operations following an unplanned interruption is redundancy. As depicted in Figure 8.6, a redundancy strategy envisions a second data center owned by the company that is fully equipped to handle system workloads—either at full production levels or at emergency levels—should the first data center become non-operational.

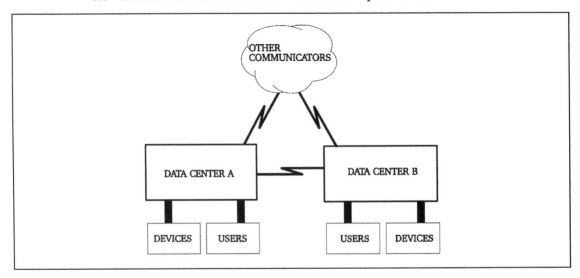

Figure 8.6 Full redundancy strategy.

In practice, switching over to this redundant data center may be accomplished within hours or even minutes following an unplanned interruption. All networks are duplicated and each site provides facilities from which end users can operate.

The expense of this option is its primary drawback. However, companies whose critical applications are time-sensitive in the extreme may have no alternatives that will meet their recovery timetable.

Commercial Hot Sites

The solution with the next highest confidence ranking is the commercial hot site strategy. Commercial systems backup centers are leased facilities where system components are installed and ready for use by companies who pay a monthly or annual subscription fee.

As depicted in Figures 8.7 and 8.8, there are two varieties of commercial hot sites. In the traditional hot site, Figure 8.7, a company subscribes to a designated facility where a specified system configuration will be maintained. A larger vendor may offer one or more alternate facilities should the primary facility to which the subscriber has been assigned be occupied or otherwise unavailable when the subscriber's disaster declaration is made and the contract is invoked.

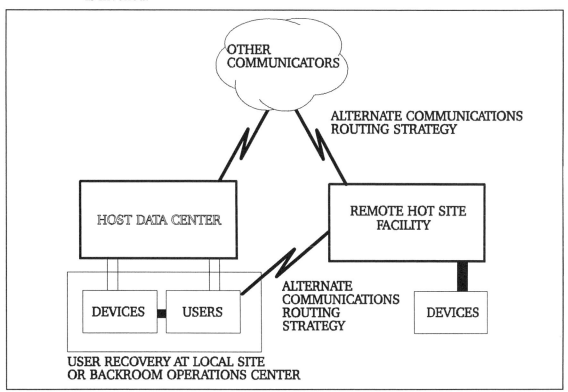

Figure 8.7 The traditional hot site backup strategy.

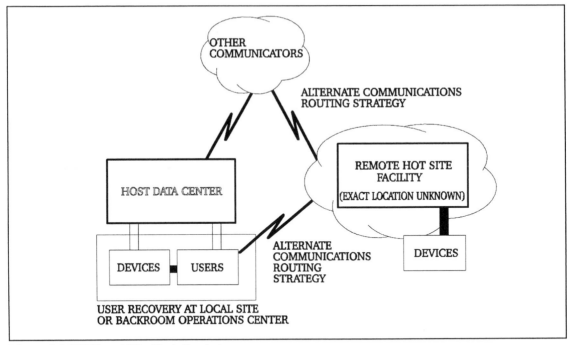

Figure 8.8 "MIPS, not Sites" hot site strategy.

Recently, a variation of this traditional hot site backup strategy has been set forth. As shown in Figure 8.8, the "MIPS, not Sites" strategy envisions the same subscriber-vendor relationship. The subscriber contracts for a certain capability for systems backup from a vendor offering multiple recovery sites. The difference between this variant and the traditional hot site arrangement is that no specific recovery location is specified in the subscription contract. The vendor agrees to recover subscriber processing from a location of its own choosing.

The differences between the two types of hot site agreements are small but potentially significant. The traditional hot site strategy affords the development of an affinity between customer recovery personnel and the vendor technicians who will aid the recovery process. Tests are conducted at the designated site and the vendor and customer are able to work together in much the same manner as they would in an actual emergency. This is not the case in the MIPS, not Sites variant.

Also, in a traditional hot site contract, the designated site can be audited to ensure that it is not leased to too many companies—oversubscribed—or underequipped to provide the promised backup capacity. This is nearly impossible to do in a MIPS, not Sites arrangement.

Finally, for companies having special equipment needs, recovery at a traditional hot site has the advantage of providing a location for storing redundant backup hardware. In the MIPS, not Sites arrangement, specialized equipment must be shipped from a central location to whatever vendor facility may be used to provide backup. This may delay recovery.

Depending on the requirements of the company, either approach may be feasible. However, companies need to be wary of signing any agreement that promises a capability for backup that cannot be demonstrated for its availability or capacity.

Shell Site

The shell site is basically a computer-ready facility equipped with power, environmental controls, and network connections but lacking any installed computer equipment. Some companies elect to convert available facility space into a shell that can be equipped and occupied if a disaster impacts the primary facility.

Commercial Shell Site

If a hot site is a leased facility with installed hardware that can be quickly configured for use in a crisis, a commercial cold site—or shell site—is a leased facility prepared for equipment installation. Figure 8.9 depicts the shell site strategy.

Like the company-owned shell site, a commercial shell facility is equipped with power, environmental controls, and network connections, but does not have installed equipment. The cost for subscribing to a commercial shell site is generally less than the cost of a hot site subscription. Equipment installed in the shell may ultimately be relocated to the subscriber's permanent facility once the facility is prepared for occupancy.

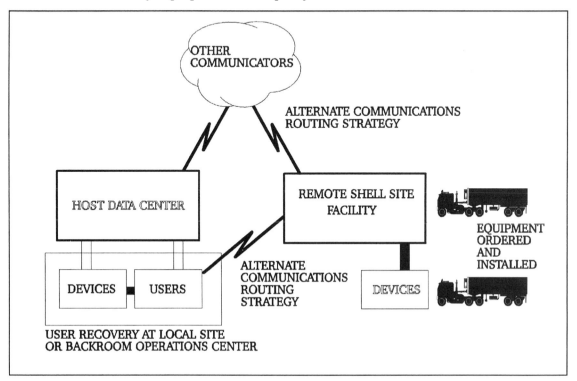

Figure 8.9 Cold or shell site backup strategy.

Shell sites are frequently offered by hot site vendors and computer-leasing companies. For hot site vendors, the cold site serves as a fallback for companies that have protracted recovery requirements. The hot site vendor may require companies to transition from the hot site into a shell facility after a fixed period of time.

Leasing companies may offer a raised floor environment for disaster recovery as a vehicle for leasing or reselling equipment to prospective clients. They promise to equip the facility with required hardware within a specified period of time should the client company experience an emergency.

If a private or commercial shell site is selected as a primary systems backup strategy, planners should be aware of the limitations of this choice. Using a shell site precludes the possibility of testing the actual implementation of the recovery system. Moreover, the recovery timeframe is hostage to how quickly new equipment can be ordered and installed in the wake of a disaster.

Service Bureaus

Service bureaus may be available to provide backup processing on an application-by-application basis. If a company determines that only certain applications are mission-critical, and a service bureau is available to provide this processing, this strategy may be a viable one.

The drawbacks of this strategy, however, are several. In general, a service bureau cannot provide a comprehensive backup capability for a company's critical systems and networks. Moreover, it is difficult or impossible to test a service bureau strategy.

Reciprocal Arrangements

Reciprocal arrangements are agreements established between two companies in which each promises to provide backup for the other in an emergency. These agreements are difficult to test and are generally considered inadequate to the task of backing up complex systems and networks.

Replacement (Low Confidence)

Replacement, often called a laissez-faire strategy, is regarded as the lowest confidence strategy for systems backup. As depicted in Figure 8.10, this approach envisions the replacement of hardware, software, and facilities on-the-fly in the aftermath of a disastrous interruption.

Not only is this strategy impossible to test, but also it will generally fail audit requirements as it is tantamount to doing nothing about the possibility of a disaster.

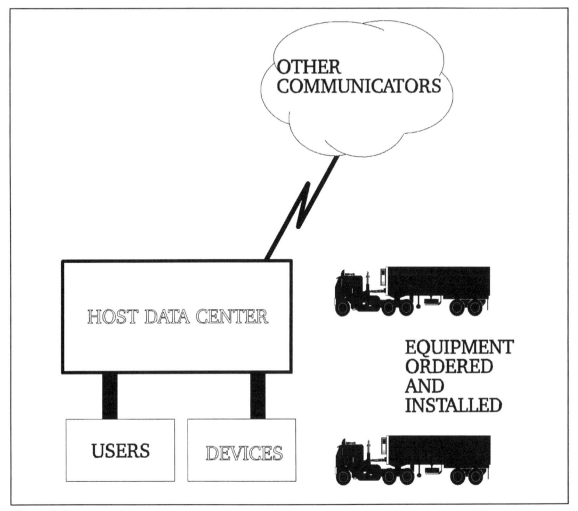

Figure 8.10 Laissez-faire strategy.

Assign Costs to Options

There is a correlation between the confidence levels associated with systems backup strategies and their costs. In general, the greater the confidence level associated with a strategy, the more the strategy will cost as part of advanced preparation for disaster. Full redundancy is the most expensive strategy and affords the highest confidence of success in a disaster recovery situation. By contrast, a laissez-faire strategy is the least expensive, but it also affords the least confidence of success.

As previously discussed, the disaster recovery planner will need to balance strategy cost with cost exposures confronting the company in the event of a disaster. Dollars spent on advanced readiness should not exceed the risk to the company expressed as potential dollar loss reduced by a percentage of likelihood of threat occurrence.

For this reason, an increasing number of companies have elected to adopt the commercial hot site strategy for systems backup. Hot sites, while expensive, are often tolerably so and these costs can be legitimated under most cost-benefit analyses. To select an appropriate hot site strategy, the following steps may be observed.

Contact Vendors

As of this writing, there are no fewer than eight established vendors of hot site services. A listing of vendors can be obtained through publications in the contingency planning field. These companies and their successful recovery stories are frequently covered in the trade press and a minimum of effort is required to identify several qualified candidates. Moreover, hardware and software vendors can often suggest a hot site services vendor if asked.

Initial vendor contacts will typically result in the visit to your site by vendor representatives. The hot site business is hotly contested and vendors will go to extremes to cultivate a positive impression in a prospective client.

Obtain Price Quotes

Having contacted vendors and received their literature, provide the minimum configuration requirements summary to each vendor and request a price estimate for a typical hot site solution.

Document Option Costs

When pricing is received from several vendors, develop a comparison sheet summarizing the offerings and pricing submitted by each vendor. By documenting option costs, the planner can compare bid pricing for a customized solution to the typical pricing originally quoted. This may serve as an effective means for estimating the additional costs associated with the application of a hot site strategy in the specific case of the company.

EVALUATE OPTIONS

Cost is only one criterion for evaluating competing options. Others are discerned from site visits, reference checks, and contract provisions.

Visit Qualified Sites

Visits should be made to all vendor sites considered in the final round of the evaluation process. Visits to the vendor site will confirm the facts of the vendor offering and also provide planners with an enhanced appreciation of how the hot site location will need to be accessed in the event of an emergency.

Check References

Vendors are prepared to provide references to prospective clients. Recognizing that these are probably the vendor's star accounts, it is still important to solicit information about the vendor's service quality, frequency and types of testing conducted, the vendor's participation in testing, and other facts about solution performance.

At least three references should be interviewed. Planners should ask to be provided references from companies in similar lines of business and/or with similar configurations to backup. Document all reference account responses carefully for later review.

Review Contracts

Review the terms of the agreement with the hot site vendor. Several policies may be specifically addressed in the agreement, as set forth in the following sections. If these policies are not addressed, find out why they are not and obtain a written statement of policy as an addendum to the agreement.

Policies on Site Subscription

Several issues need to be resolved up-front in hot site contract negotiation. These include the following:

- Identify what the vendor's contractual position is on service subscription.
- Determine whether a specific site is designated for testing and backup.
- Determine whether the site is reserved only for subscribers or if other customers may be granted access in an emergency.
- Identify how much testing time is provided each year and at what cost, and what support can be expected from vendor technical personnel to support testing.

The answers to these questions may provide tangible differentiators that will enable a clearer decision to be made between vendors with comparable service offerings.

Policies on Disaster Declaration

Determine how the vendor handles disaster declarations. Many vendors charge a disaster declaration fee to "discourage frivolous use" of the facility. Some do not. Also, some require a certain advance notice or provide facilities to customers on a first come, first served basis. Determining the position of each vendor on these issues may help to differentiate vendor service offerings.

Policies on Facility Loading

As a subscription service, hot sites are typically leased to several companies concurrently. This is done to reduce the costs of the services to all subscribers, to earn the vendor maximum profit, and to acquire revenue for use in upgrading services to keep pace with the latest technologies.

The planner needs to obtain a clear statement from the vendor regarding the number of companies booked to the recovery platform that is being offered to the planner's company. Watch for any indication that the vendor seeks to share partitions within the same platform between multiple companies or with its own production systems. Seek a dedicated system solution as a rule-of-thumb.

The vendor should also be asked how many companies in the same geographical area as the planner's company are booked to the same system. This may be a factor if a regional disaster impacts several companies in the same area.

List of Other Services Commonly Provided by Commercial Recovery Vendors

Planners should ask vendors about other services provided by the company. Services of interest may include:

- Support for non-mainframe system recovery
- Shell site services
- Plan testing and maintenance services
- Telecommunications and network recovery services
- Backroom operations center services for user recovery

Solicit Bids

Once the field of choices has been narrowed, solicit formal bids from each vendor. It is wise to instruct the vendors to provide their best pricing in their bid. In this way, the unfair and distracting process of last-minute pricing adjustments can be avoided.

DOCUMENT PREFERRED STRATEGY

Whether a hot site or other systems backup strategy is preferred, the option selected and the rationale for its selection—including vendor bid analysis—should be scrupulously documented. This will enable the planner to defend against any subsequent management challenges regarding the strategy selection and bid evaluation procedures.

Once the strategy is identified, it can be presented to management for approval. When approved, the planner will set about identifying the actual tasks that will need to be accomplished to realize the goals of the strategy in an actual emergency. These tasks and their related procedures will comprise a section of the physical planning document.

SUMMARY

System recovery strategies come in all varieties of expense and complexity. The appropriate strategy for any given company is the strategy which will provide for the restoration of minimally acceptable functionality, timeframe, and cost.

It should be kept in mind that system recovery strategies are not formulated in a vacuum. They are inexorably linked to end-user and network recovery strategies and should be reviewed whenever these other strategies are fully articulated.

Checklist for Chapter

1. Inventory target system hardware and software.

2. Determine minimum equipment configurations and document requirements.

3. Identify data communications requirements.

4. Identify end-user requirements.

5. Evaluate system recovery options. Evaluate and select vendors to bid. Assess bids and prepare systems backup strategy proposal with costs.

Network and Communications Recovery Planning

SETTING THE SCOPE OF NETWORK RECOVERY

The shift from centralized to decentralized computing in the 1980s gave rise to a 1990s trend to reconnect dispersed islands of automation within the company via local and wide-area networks. Networks are utilized to transport computer, voice, and video data between corporate offices. Networks—as data transport technologies—facilitate primary business functions as much as data processing technologies. Their recovery is critical, therefore, to the successful recovery of the business from a disaster.

In one sense, networks provide the glue of the recovery plan. Strategies articulated for system and end-user recovery will not succeed if provisions are not made to recover the networks that link host systems to users, or systems and users to external entities including customers, government agencies, financial institutions, and so forth. Without the recovery of key network facilities, the disaster recovery plan cannot succeed.

Networks themselves can also be the location of a disaster and their restoration the objective of disaster recovery planning. Wide-area networks can be interdicted in many locations, as depicted in Figure 9.1.

A typical network access path involves equipment operated from the company site, such as a private branch exchange (PBX), router, fabric switch, bridge, multiplexer or concentrator, modem, or DSU/CSU. From this equipment, lines extend to the local service office or central office (CO) of the local telephone company. Local telcos typically provide high-speed access to the wide-area network services provider or long-distance carrier offices, these offices provide a gateway into the wide-area network comprised of public and private communications facilities and switches.

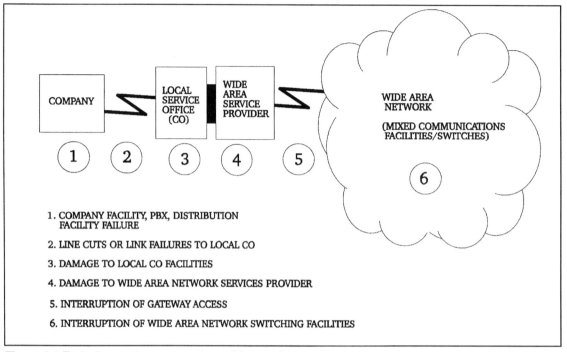

Figure 9.1 Typical network access and possible interdiction points.

Closer observation of this simplified network access path reveals several vulnerabilities. It can be interrupted by equipment failures, line cuts, CO fires, carrier office fires, gateway failures, and switch or communications facility failures, just to name a few potential problems.

To most companies, the loss of communications access to other offices, to customers, or to external entities will result in a significant loss of business. For this reason, a phase of the disaster recovery planning project is dedicated to network and communications recovery planning. Figure 9.2 describes some of the main activities of this planning phase.

As with systems backup, network backup planning seeks first to discern the network configurations that support key business functions. Then, an effort is made to develop a consolidated network plan, or minimum service requirement, that will support critical applications acceptably in an emergency. Options for providing backup and recovery capabilities for key network assets are assessed. Finally, a mix of services and products is selected for inclusion in a strategy for network recovery.

THE DISASTER RECOVERY PROJECT
NETWORK BACKUP

MAJOR ACTIVITIES

Develop Consolidated Network Plan

Select Re-Routing Option

Evaluate Vendor Services

Document Strategy

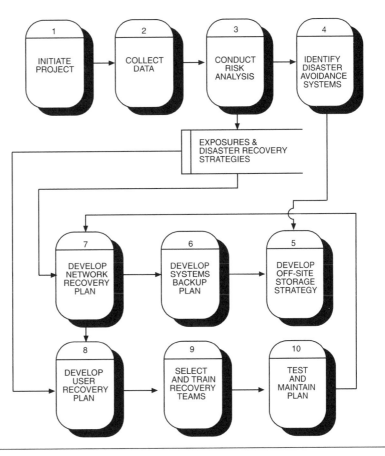

Figure 9.2 Network backup phase.

Time Requirement

As stated previously, the success of user and system recovery strategies is contingent upon the successful implementation of new network connections in a timely way. The timing of network restoration tasks is closely linked to system and user recovery tasks in a typical disaster recovery plan. The network must be operational by the time that systems and users are restored. Thus, network restoration efforts commence with the onset of a disaster and run concurrently with other recovery efforts.

Only with advanced planning can network restoration be accomplished within the timeframe for overall recovery set forth for the plan. Many companies have learned through testing and actual disaster recovery experience that failure to plan and test network recovery strategies can delay implementation of the company disaster recovery capability for many hours or even days following completion of system and user recovery tasks.

Inventory of Critical Business Functions

Defining network restoration requirements begins with the examination of critical business functions. Critical functions were defined as part of the risk analysis phase of the disaster recovery planning project. Details of the functions are recorded in the business function survey and subsequent analysis worksheets.

If work has already been completed on the systems backup strategy, it is possible that the inventory of critical business functions has already been performed. This does not mean that all critical functions are supported by the systems backup plan. While the systems backup strategy may identify the preponderance of network recovery requirements that need to be provided in the network backup strategy, it may not include those functions that have no data processing requirements. For example, a critical telemarketing function may only be identified in the user recovery strategy-setting process, if this function has no prerequisite data processing requirements.

In short, all business functions need to be reexamined to determine their network support requirements. Other plan strategies may provide important inputs to this process, but they do not replace the need to perform a separate and complete audit.

Requisite Communications Capabilities

Once critical business functions have been inventoried, their requisite communications capabilities need to be documented. This task is far-reaching and includes the documentation of communications hardware on the company premises, links to and locations of local COs, wide-area network gateway service descriptions, and wide-area private network descriptions.

Internal Equipment

Prepare an inventory of on-premise communications equipment, including company PBXs, MUXes, DSU/CSUs, modems, WAN routers and bridges, channel extenders, and so forth. Associate equipment with the business functions that have network communications requirements and describe the role played by the equipment.

Local Company Office (CO) Services

Identify the local telecommunications CO or switching center serving the company's local telecommunications needs. Identify any alternative COs that might be utilized should the primary facility become unavailable.

It is also important to document telecommunications links—including lines, trunks, or digital facilities—that connect company voice and data networks to the local CO. Special services—such as voice mail services, call accounting services, and others—provided by the local CO should also be documented.

Wide-area Network Gateway Services

Investigate and document gateway facilities provided between the local CO and wide-area carrier offices. Determine how these services will be provided in the event of a local CO outage.

If the company uses wide-area networks based upon public-switched networks, identify how access to the public-switched network is provided and how access will be restored if an interruption occurs.

Wide-area Private Network Services

Document private networks employed by the company. Identify interexchange carrier points-of-presence that provide access to the wide-area network. In documenting network topology, be sure to document such special capabilities as satellite links, T–1 and other facilities, subscription services, and digital radio.

Satellite

Some companies, often for reasons related to regional geography, utilize satellite links to handle important voice and data communications. If this is the case, the planner should document the satellite communications link and configuration.

T–1 and Digital Facilities

For reasons of bandwidth capacity, many companies lease fractional T–1, full T–1, and T–3 facilities to establish wide-area internetworks or to operate channel-attached devices at remote locations via channel extension. The use of these facilities needs to be identified and their purpose clearly understood in order to plan for recovery of business functions that depend on these links.

Subscription Services

Some regions of the USA are served by specialized subscription services for wide-area internetworking. Such services are marketed to the company that does not wish to manage a private internetwork and prefers to lease a custom network alternative from a third party. Planners need to document the usage of these services where appropriate and identify interdiction threats and recovery requirements associated with them.

Digital Radio

Packet radio continues to find use in some areas where land-based links and satellite services are unavailable or insufficient to the company's requirements. If digital radio is employed, the planner should document equipment, access methods, network characteristics, and vendor-provided gateway functions.

DEFINE A MINIMUM SERVICE REQUIREMENT

Once business functions have been reviewed for their network support requirements in normal operation, the next step is to determine what consolidations can be made to meet the needs of the business operating in a disaster recovery context. This begins with a review of subject networks.

Review Subject Networks

High-level descriptions should be prepared for all networks employed by the company. This effort is facilitated by the inventories and research conducted in the preceding task. The networks will be associated with critical applications once the descriptions are complete.

At a minimum, three network descriptions should be prepared:

- Local Voice/Data
- Wide-area Voice/Data
- Private Networks

Local Voice/Data

The local voice/data network description covers network topology, physical media, and communications equipment used to facilitate voice and data communications within a local area. The term *local area* may be interpreted to mean within a building, within a campus, within a metropolitan area, within an area code, or within the state. The point is that traffic over this network is bounded by the local telco central office or switching node that the telephone company uses to define an area code. Traffic over this network is generally free of toll charges.

Wide-area Voice/Data

A second network requiring description is a wide-area network utilizing interexchange carrier services. The company uses a wide-area network when making long-distance telephone calls. Similarly, extensive dial-up networks connecting corporate facilities outside the local area may be regarded as wide-area networks. Toll charges apply to communications made in this network as the public telecommunications switching network is utilized.

Private Networks

Private networks are wide-area networks configured and managed by the company. In some cases, leased lines and other digital services and facilities may be acquired from third-party providers. The point is that the network is dedicated to the company's exclusive use and is configured to meet company needs precisely.

This network may also be viewed to include hybrid networks—that is, networks configured by the company and managed by the company using on-premise equipment, by a public telecommunications company using its equipment, or by some combination of both. There is a growing trend toward hybrid networking as value-added networks, frame-relay networks, and ATM network facilities become more available and less costly.

As with the other network descriptions, provide an accounting of the network's topology, equipment involved, physical links, and traffic type. This will facilitate the identification of the business's functional dependence on the network.

Identify Critical Applications

Using surveys and analytical documents created in the risk analysis phase, identify critical business functions and their related applications and equipment requirements. Determine on an application-by-application basis which networks are involved to support which critical functions. Create a list of network requirements for critical functions.

Identify Required Peripherals

Review equipment lists submitted by business work units. Determine what are the peripherals, including telephone handsets, fax machines, PC modems, and other devices that have been identified as critical for performing the critical tasks of the work unit. Add this equipment to the requirements listing.

Explore Opportunities for Integration/Consolidation

Given that some business functions may be less sensitive to interruption than others, their related network and data processing resource requirements may not need to be recovered in the initial period following a disaster. Other business functions may be able to be handled by fewer staff, from alternate locations, or on a shift schedule basis, possibly impacting bandwidth requirements.

This situation suggests that it may be possible to integrate or consolidate network requirements so that a recovery network configuration with less capacity than the company production network can be defined. The benefit of defining a somewhat less complex network configuration is that it can be implemented more quickly and placed into operation more readily.

Technical assistance will be required to determine what the new access and bandwidth requirements are for the disaster recovery network. Planners should invite vendors and in-house network experts to participate in the definition of an emergency backup network that will handle the load of critical business functions in an emergency.

Explore Opportunities for Traffic Rerouting

In any disaster, there is the likelihood that some communications traffic will need to be rerouted. For example, if user work areas are destroyed in a fire and users are forced to work from another office, telephones, fax machines, modems, and other voice and data connections will need to be rerouted to the new work area.

Dial backup is the strategy familiar to many disaster recovery planners for accomplishing this task. As depicted in Figure 9.3, traditional dial backup provides a means for rerouting voice and data communications should a communicator shift location.

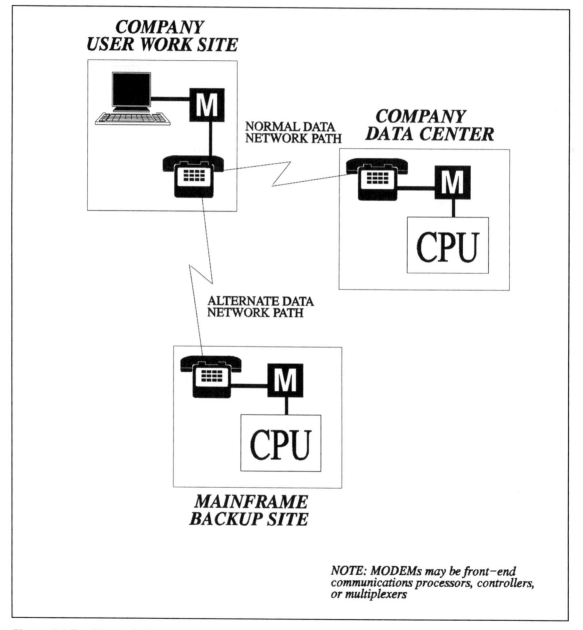

Figure 9.3 Traditional dial backup.

A similar approach has been successfully applied in the area of channel extension. Channel extenders enable mainframe peripherals that are normally cabled to the backplane of the mainframe to be placed at remote locations. Communication between the mainframe channel and the remote device is accomplished through a broadband medium such as fiber optic cable, serial or wide-area network links, or digital communications facilities. If the data center at a company is rendered unusable, as depicted in Figure 9.4, and a mainframe backup plan is implemented, channel extension may be used to enable users to remain in place, but to communicate with the remote host over a wide-area link.

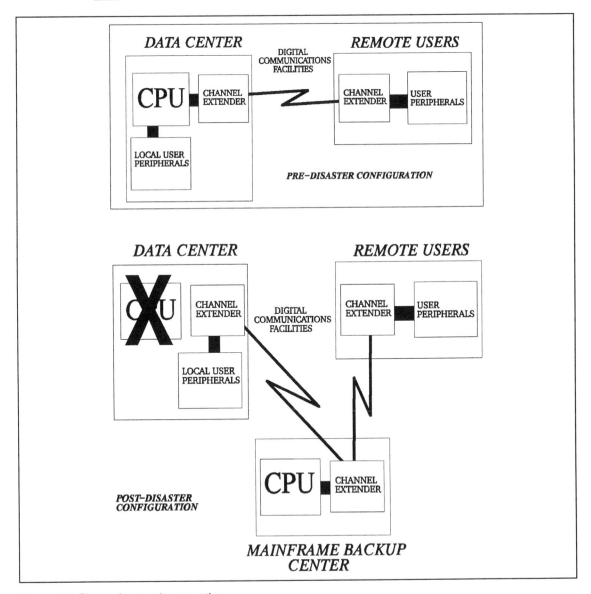

Figure 9.4 Channel extension rerouting.

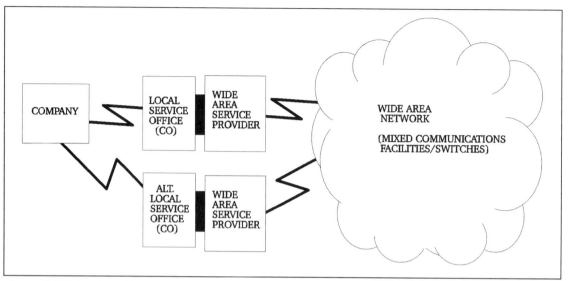

Figure 9.5 Network routing alternatives: Network access through an alternate CO.

Rerouting the path for communications between the host and remote, channel extended, devices may be viewed as a variation of the dial backup strategy. For such a strategy to work, however, the requirement for an alternate path needs to be known in advance and plans need to be put in place for establishing the new communications route when needed.

Rerouting is also an important consideration when looking at risks to networks themselves. As indicated at the outset of this chapter, network access can be interdicted at several points along the access route, resulting in network downtime that might impact the company adversely.

One scenario envisions the interruption of the functionality of the local telephone company CO. A fire at the CO, for example, could prevent the operation of local voice and data networks and also constrain access to the wide-area network.

An alternative, depicted in Figure 9.5, is to establish an alternative traffic path to another local CO in the company's vicinity. This redundancy may be costly to realize, but it can afford substantial risk reduction that justifies its cost for a company that is dependent on local voice and data networking.

A variation on the routing redundancy strategy is depicted in Figure 9.6. In some cases, companies do not wish to depend upon a single CO or interexchange carrier point-of-presence to gain access to the wide-area network. If an alternative wide-area network access facility can be reached via an alternative route, implementing this topology may reduce risks to critical corporate networks.

These strategies will surmount a number of possible access interdictions resulting from causes located in the physical link or CO facility. Upon review, planners may find that they are satisfied with the resilience of the local loop or they may determine that it is too expensive to establish a redundant path. Concern about the issue may remain nonetheless.

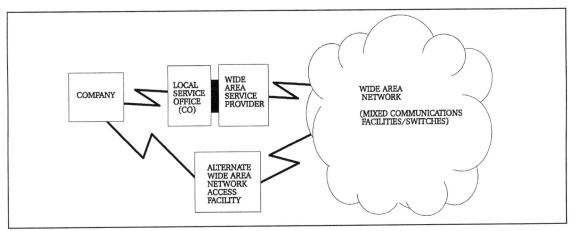

Figure 9.6 Network routing alternatives: Access through an alternate facility.

To cope with the possibility of a link interruption, media other than analog and digital land lines may be considered. These alternatives are summarized in Figure 9.7.

Using an alternative media, such as microwave, may provide a less expensive means to implement a rerouting strategy should a link failure or CO outage occur.

With traffic rerouting strategies explored and documented, another key element in the formulation of an emergency network backup strategy is complete. The bandwidth requirements of the emergency network, however, cannot be determined without reference to its actual use.

Determine Usage Parameters

Usage may determine the type and capacity of link media used in the recovery network. Usage characteristics, therefore, must be postulated when defining the recovery network.

Skeleton User Population/Shift Scheduling

As discussed in Chapter 8, usage of systems and networks in a corporate recovery situation may be very different than in production systems and networks. Two factors may account for this difference.

For one, the user population in a recovery situation may be much smaller than in normal operations. Skeleton crews may be used to perform the functions of entire work units. Fewer personnel may require voice and data communications access to provide key business functions at minimally acceptable levels. Thus, less bandwidth or less connectivity may be required.

Even when work units remain virtually the same in size in the recovery scenario, network demands may be reduced. If applications are available for use on a scheduled basis, work may be conducted in shifts rather than having all users working concurrently. This, too, may reduce bandwidth and connectivity requirements in elements of the recovery network. These usage parameters need to be documented for reference during the formulation of an overall network recovery strategy.

ANALOG AND DIGITAL LINES

Traditional telephone networks use analog technologies with a recommended maximum speed of 19,200 bps

Digital networks offer greater range of transmission speeds and much lower error rates than analog

CELLULAR TECHNOLOGIES

Cellular radio uses low-power radio transmitters in a honeycomb configuration to permit reuse of a frequency multiple times in a small area -- useful in cases of local line failures

DIGITAL RADIO

Digital radio is an omni-directional medium without the alignment requirements of terrestrial microwave or satellite communications

Interference may result from line-of-site obstructions, though this problem may be defeated through use of store-and-forward repeaters that receive data "packets" in short burst transmissions

MICROWAVE

Uses a parabolic dish to focus a narrow beam for high-speed, line-of-sight data transmission

Subject to propagation delays over long distances

SATELLITE

Similar to terrestrial microwave, but uses a satellite as a relay antenna

Low cost option, but propagation delays present challenges for digital communications error and flow control

Figure 9.7 Media restoration alternatives.

Data Communications Requirements

Another determinant of the recovery network is the data communications requirements of external entities. Lockbox vendors, payroll vendors, off-site data entry contractors and others may play a role in day-to-day business operations or only in a recovery scenario. These entities may require access and bandwidth on an ongoing basis or at certain times during the processing schedule. Planners need to consider the data communications requirements of external entities as a parameter in emergency network definition.

Shift Scheduling

The usage requirements of users, applications, and external entities should be gathered and represented on a schedule for planning purposes. The schedule may be delineated on a shift/basis or other basis, showing peak utilization requirements and the times of day that they will occur.

DOCUMENT A MINIMUM EQUIPMENT CONFIGURATION

With the research conducted in the preceding section, it should be possible to define a minimum equipment configuration for the recovery network. This configuration can then be used to shop for the products and services that will be required to actualize the network recovery capability.

Hardware Requirements

Document the hardware requirements for the emergency network. Include all critical communications devices, links, and switches. This inventory can be refined as the network definition proceeds.

Software Requirements

Note all software requirements for communications devices. Include operating system software and custom node lists for routers, channel extender configurations, PBX software, communications software for use with modems, terminal emulation packages, and so forth.

In addition, note any changes that will be required to application software to facilitate their communication with external entities or other company nodes in the wake of a network reroute, media change, or other deviation from normal operation. Some applications, for example, may be sensitive to time delays that may accrue to the use of satellites or ground transmissions over 250 miles. These applications may need to be tuned to work well with the characteristics of the recovery network.

Assumptions

List assumptions that are being made about the performance characteristics of the new network. Be sure to include the numbers of users and schedules for usage that are anticipated. Provide a list of critical hardware that must be in place for the network to be realized.

EXAMINE RECOVERY OPTIONS

Strategy options for network recovery are few in number. Comprehensive, packaged solutions are rare and generally very expensive. Typically, the network recovery strategy will include a mix of services from vendors and internal preparations.

Options

The options in the network recovery strategy are the same as for systems recovery. A strategy of full redundancy is the most expensive option and affords the highest confidence of success. A laissez-faire strategy—taking no preparatory steps—is the least expensive with the lowest confidence of success. For a company that uses networks to provide critical business functions, the best strategy will probably reside between these two extremes.

Redundancy

Clearly, any viable network recovery strategy will entail some redundancy. Alternate routing paths will be identified or replacement hardware will be reserved. Not all redundant aspects of the strategy need to be purchased by the company in advance of the disaster, of course. Some vended services are available to provide redundant capabilities shortly after notification of an outage.

Alternate Network Services

At its simplest, redundancy suggests that an alternate network can be used if a primary network fails. In practice, this scenario might involve the changeover of a company's private wide-area network to a public or hybrid network vendor's net should an interruption occur. Most metropolitan areas are served by a number of value-added network vendors who might be able to map out a strategy for switching to an alternate network service on a notification basis.

Check with the local telephone company or review trade periodicals for the telecommunications industry to identify potential vendors. Ask for a consultation to determine the cost for this option.

On-Demand Rerouting

Most major interexchange carriers, and even some local telcos, offer an on-demand rerouting service. This service is typically made available on a subscription basis and is billed as part of the monthly company telecommunications bill. Contact your service providers for a consultation.

Notification Rerouting

In addition to on-demand rerouting, some carriers and value-added network vendors offer rerouting services on a fee basis. In other words, no subscription charges accrue. To obtain the service, the company must contact the vendor when the service is required and wait the standard interval for the new network links to be implemented. Notification rerouting may be difficult to obtain, however, in regional disaster situations. In these instances, many companies may contact the vendor concurrently, resulting in a queue for service. Contact a vendor or carrier for more information.

Repair and Replacement

Of course, repair and replacement of faulty network components is always a recovery option. Arrangements can be made with an equipment supplier to provide replacement components on a priority basis. In some cases, critical hardware may be able to be stored at the company site by vendors who need storage for their products. In every case, however, the implementation of a strategy based on replacement and repair has the associated drawback of being difficult to test and maintain.

Assign Costs to Options

With the component elements of a network recovery plan identified, options should be considered and their costs identified. This is a two-stage process involving, first, vendor contacts, and second the solicitation of price quotes.

Contact Vendors

To identify vendors of network recovery products and services, planners can attend disaster recovery conferences, consult network trade journals, ask peers in other companies, talk with telecommunications providers, and contact professional network consultants for advice. Several vendors should be contacted for every element solicited from vended sources.

Obtain Price Quotes

Meet with vendors and set out the network recovery strategy, describing in detail the part that the vendor's product or service needs to play. The vendor may be able to identify options that the planner did not consider previously.

Obtain a description of the product or service of the vendor and solicit a price quote. Add this to the network recovery strategy documentation for later review.

Document Option Costs

Once pricing has been obtained from vendors, the costs associated with various network recovery options need to be documented. For comparison purposes, it may be beneficial to document the costs entailed in a strategy of full redundancy, as well.

EVALUATE OPTIONS

With several options identified and priced, a formal evaluation process should be conducted. The results of this process will be a strategy for recommendation to senior management. Ensure that the evaluation criteria are clearly defined and that the benefits and limitations of each option are presented for ease of understanding.

Check References

Obtain lists from each vendor of three current customers and contact each customer reference provided. Document the customer's experience with the vendor and identify commonalties and differences between the customer's requirements and those of the company for which the network recovery strategy is being formulated.

Review Contracts

Ensure that all contracts for services are reviewed by competent technical and legal authorities. Look for hidden implementation fees or installation charges. Determine whether the vendor assumes any liability if the product or service being promised fails to perform as expected. Ask for testing to be included in the arrangement.

Solicit Bids

When an option is selected, solicit bids for all of its component products and services. Ensure that the bid pricing provided will not expire before the presentation of the strategy to senior management for approval.

DOCUMENT PREFERRED STRATEGY

Once the preferred strategy has been documented, identify tasks required for implementing the recovery network. These high-level tasks will be broken into subtasks and procedures once the network recovery strategy is presented and approved by management for implementation.

SUMMARY

Networks are the paradigm of modern enterprise computing. Networks are also, in themselves, potential points of vulnerability for a business. Protective measures must be taken to prevent avoidable disasters and to recover expediently from disasters which cannot be avoided. Their restoration is critical to business continuity.

To a great degree, the survivability of a network is linked to its design. Networks featuring redundancy of key nodal elements or architected with alternative links and message paths are durable and less prone to interruption than those lacking these design elements.

However, both system and end-user disasters will have network recovery requirements. Even if the network is not affected by the disaster which disables the data center or the end-user department, it will need to respond to these situations. Once formalized, it is important to review the network recovery strategy within the context of system and end-user recovery requirements.

Checklist for Chapter

1. Identify locations and type of telephone and network service to company facilities.

2. Identify communications requirements for critical business functions.

3. Define a minimum service requirement.

4. Document a minimum equipment configuration.

5. Evaluate network recovery options. Select a strategy and identify costs. Solicit and evaluate vendor bids. Document the preferred strategy for presentation to management.

10

End User Recovery Planning

SCOPE OF USER RECOVERY PLANNING

Disaster recovery planning aims at recovering mission-critical business functions in the wake of unplanned interruptions. Traditional disaster recovery planning focused on recovery of the data center and applications residing on a mainframe host. Little attention was devoted to networks, distributed systems, or end users.

Today, the proliferation of departmental computing, desktop workstations, and local area networks has led to a complex problem for traditional planning. Applications based on distributed architectures, heterogeneous operating systems, and client/server technologies have added to the problem.

The facts are simple. An increasing volume of critical data resides in user work areas. Moreover, user work areas are more likely to be the site of a disaster than are data centers with their secure, environmentally controlled, routinely backed up, and power-protected systems.

Even more important are the personnel who work in departmental offices and cubicle work spaces to make corporate computers and networks purposeful. The first priority of disaster recovery planning is personnel safety.

For these reasons, the scope of disaster recovery planning must include end user recovery planning. Following a disaster, users must know when and where to report to work. They must have a *backroom operations center* identified in advance and equipped with all of the critical supplies and equipment required to provide critical business functions.

Figure 10.1 summarizes the key activities involved in development of the user recovery plan, the seventh phase of the disaster recovery planning project. This chapter discusses the user recovery planning tasks in detail.

THE DISASTER RECOVERY PROJECT
USER RECOVERY PLAN

MAJOR ACTIVITIES

Identify User Work Requirements

Identify Supplies and Equipment Requirements

Evaluate Vendors

Develop Logistics Plan

Select and Equip Backroom
Operations Center

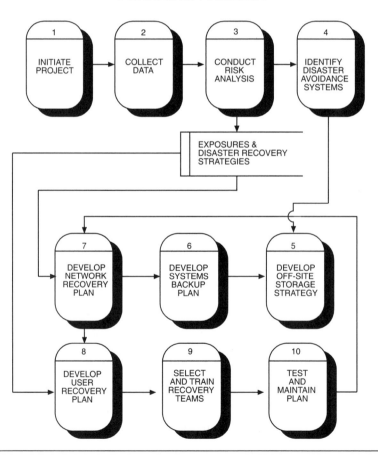

Figure 10.1 User recover planning phase.

Resumption of Critical Business Operations

The resumption of critical business operations requires the implementation of a plan that will provide company personnel—the users of corporate information systems and networks—with everything they need to meet their work requirements. This effort has several components.

Users must have a location from which to work. The facility need not be spacious or well-appointed, but it must provide a comfortable surrounding where useful work can be performed. The user recovery location must be equipped with the supplies and office machinery identified in the Business Function Survey. Moreover, it should be established with overnight mail services and the Postal Service as a forwarding destination for correspondence directed to the evacuated corporate facility.

Systems Recovery

In addition to supplies, furnishings, office machinery, and mail, end users need to have access to system resources used to access mission-critical applications. The systems recovery plan should provide for the restoration of all shared system resources, but the acquisition and installation of end-user terminals or workstations may become part of the user recovery plan.

Network Recovery

End users also need to have access to voice and data communications capabilities to do work with recovered systems hundreds of miles away. There is at least one synergy between user and network recovery planning: network recovery is facilitated if the location of the user recovery center is known in advance. A predesignated location for user recovery facilitates the implementation of network rerouting strategies.

Recovery Personnel

Who will be included among the personnel responsible for sustaining critical business operations from the user recovery location? There are no hard and fast answers to this question. Each department manager will need to identify the personnel who are appropriate for inclusion in the backroom operations center staff.

For most business recovery functions, fewer staff will be required than are needed in normal operations. It is important that the staff selected to participate in the user recovery center possess strong familiarity with the work unit's procedures and experience across the entire workflow of the unit.

Technical/Managerial Roles

Every work unit will require representatives with managerial and technical skills. These personnel will be responsible for assisting in the preparation of the new work area following a disaster and will participate in the maintenance of the disaster recovery plan as it pertains to their work unit.

User Roles

In addition to technical and managerial personnel, the work unit will also be represented by personnel who provide the knowledge and skills to accomplish the business function performed by the work unit. The work of these personnel constitutes the actual recovery of the business. Customers will interact with these personnel, systems will be operated by them, and networks will be made purposeful by them.

Users will also be the primary source of information about the performance of recovered systems and networks. Their requirements will be used to fine-tune resource allocations, processing and shift schedules, and system and network configurations.

BACKROOM OPERATIONS CENTER STRATEGIES

The previous conditions presuppose that users will relocate to an alternate facility or backroom operations center in the event of an emergency. Figure 10.2 provides a set of criteria planners can use to think about and evaluate backroom operations center options.

As indicated in Figure 10.2, an effective backroom operations strategy entails a facility sufficiently distant from the normal production facility to prevent it from being impacted by the same disaster that renders production facilities uninhabitable. The facilities must offer adequate space, furniture, office equipment, and must be accessible via commercial transportation. The facility must offer appropriate capabilities to ensure the safety and security of personnel. It must be equipped with utilities, environmental controls, and automation resources, including voice and data network requirements.

Local Option

In light of the criteria previously advanced, a backroom operations center—such as another facility operated by the company—may provide adequate capabilities to support emergency operations, but it may be exposed to regional disaster potentials that may also affect the company such as earthquakes, hurricanes, regional power or telephone outages, and so forth.

Remote Option

Remotely located facilities have the advantage of being insulated from regional disasters, but may pose logistical difficulties and impose greater costs. For example, the cost for transporting users to perform work at a remote location is greater than a local facility option. Moreover, establishing an alternate routing path for corporate networks to a remote location may entail greater costs.

Laissez-Faire

In light of the costs and benefits of both the local and remote options, some companies elect to establish no recovery facility at all, but instead designate a team that will search for available office space in the wake of a disaster. Depending on the complexity of work unit requirements, this strategy may be workable. However, it obviates the advantages of defining a location for user recovery in advance—namely, that it enables users to be directly involved in plan maintenance and testing and that it allows planners to put concrete network rerouting strategies into their overall network recovery plan.

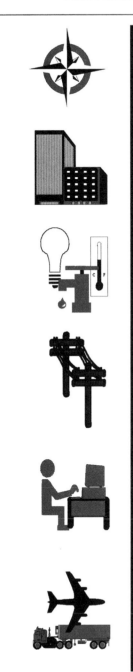

PROXIMITY

Safe distance from regional catastrophes?

FACILITIES

Provide adequate space to support work?

Disaster prevention & security systems adequate?

UTILITIES & ENVIRONMENTALS

Meets requirements of automation resources and workers?

NETWORK ACCESS

Prepared for voice & data communication requirements?

Alternative WAN access paths available?

EQUIPMENT AVAILABILITY

Office equipment and furnishings provided?

Local vendors identified?

TRANSPORTATION & LOGISTICS

Convenient to commercial transportation?

Vendors advised of new location?

Figure 10.2 Criteria for evaluating backroom operations centers.

LOGISTICS AND PLANNING

Whatever the strategy adopted for establishing a user recovery center, the user recovery plan must provide an approach for meeting the logistical requirements of the work unit recovery. These requirements may be categorized as site preparation, notification of recovery personnel and users, transportation, shift scheduling, and food and shelter.

Site Preparation

Preparing a site for use by dislocated work unit personnel entails many of the same considerations involved in any office move. The difference is that relocation to the new facility must occur in a much more abbreviated timeframe than is normally available for a nonemergency office move. Figure 10.3 provides a rudimentary checklist for practical requirements that will need to be addressed in site preparation.

Environment

The environment of the recovery site needs to offer the same basic amenities as any business office. These include adequate work space, air conditioning and heating, lighting, noise abatement, lavatories and water, electrical outlets, local telephone service, fire alarms, and security systems.

Hardware

Workstations and other computing and networking equipment should be identified in advance and delivered and installed as part of site preparation. Preparations should be sufficiently detailed to have identified and provided for cables, connectors, and other peripheral components.

Arrangements may be made with equipment vendors or leasing companies to facilitate these requirements. Also, the company may designate personnel from its own technical staff to oversee the restoration of computer systems and networks at the user relocation facility.

Software

All software required for work, including data retrieved from off-site storage, must be installed and tested on the systems in the user relocation facility. If it is necessary to install new local area networks or multiuser system platforms, per the minimum equipment configuration articulated in the systems recovery plan, the implications of this action should be considered. Any systems that deviate substantially from equipment or applications with which the users are familiar may entail a training or user oversight requirement.

Supplies

Supplies and materials required for work must be made available. Preprinted forms, paper, writing implements, and a host of other products routinely used in the office should be identified and vendors should be identified who can supply the needed supplies on an overnight basis.

USER RECOVERY CENTER REQUIREMENTS CHECKLIST

FACILITIES

LOCATION: _____

PROPERTY MGT CO: _____
CONTACT NAME: _____
TELEPHONE: _____
EMERGENCY TELEPHONE: _____

SQUARE FT: _____
WATER? _____
A.C./HEAT? _____
POWER? _____
CENTREX/PBX? _____
120V OUTLETS: _____
220V OUTLETS: _____

FURNISHINGS & OFFICE EQUIPMENT

SIZE OF RECOVERY STAFF: ☐
DESKS/CHAIRS: ☐
TEMPORARY FILING CABINETS: ☐
PHOTOCOPY MACHINES: ☐
MICROFORM READER/PRINTER: ☐
OPTICAL DISC READER/PRINTER: ☐
IMAGING WORKSTATIONS: ☐
POSTAGE METERING EQUIPMENT: ☐
COFFEE MACHINE SERVICE: ☐
RESTROOM FACILITIES: ☐

REMARKS

_____
_____
_____
_____
_____
_____
_____
_____
_____

SYSTEMS

NUMBER OF TERMINALS:
NUMBER OF MICROCOMPUTERS:
NUMBER OF PRINTERS:
NUMBER OF LAN WORKSTATIONS:
MINICOMPUTER?
ITEM SORTERS/OTHER DEVICES:
UNINTERRUPTIBLE POWER SUPPLY?
RS232 CABLES/SWITCHES:
CENTRONIX/IBM PARALLEL CABLES:
CABLE CONNECTORS, ETC.:

DESCRIPTION/MODEL/TYPE

Attach systems layout diagrams, cabling specifications, LAN/minicomputer boot procedures, and additional documentation that will aid in the speedy implementation of user recovery systems at the user recovery facility.

Figure 10.3 User recovery center requirements checklist.

SYSTEMS, CONT.

SOFTWARE NAME, VER, RELEASE

# COPIES MICRO OS:	
# COPIES LAN OS:	
MINICOMPUTER OS:	
OFF-THE-SHELF APPLICATIONS:	
(List number of copies and software make, version, release)	
MAINFRAME COMM/INTERFACE:	

Identify and list custom software products and data backups required from off-site storage.

COMMUNICATIONS

DESCRIPTION

# MODEMS& SPEEDS:	
# FAX MACHINES:	
# MULTIPLEXERS:	
# COMMUNICATIONS CONTROLLERS:	
# CHANNEL EXTENDERS:	
# DEDICATED LINES:	
PBX/CENTREX?	
SPECIAL WAN ACCESS FACILITIES:	
AUTOMATIC CALL REROUTING?	
CABLE/JACKS:	

Attach network layout diagrams, cabling specifications, installation procedures, and additional documentation that will aid in the speedy implementation of user voice and data communications networks at the user recovery facility.

GENERAL VOICE TELEPHONE #: _____

FAX TELEPHONE #: _____

HOT SITE HELP DESK #: _____

Figure 10.3 User recovery center requirements checklist. (Continued)

NETWORKS, CONT.

CONTACT/PHONE

HOT SITE FACILITY:	
LONG DISTANCE SERVER:	
LOCAL CO:	
MAINFRAME DIAL-UP 1:	
MAINFRAME DIAL-UP 2:	
SPECIAL WAN ACCESS SERVER:	
OTHER COMPANY USER FACILITIES:	
(List voice and data lines for other corporate network nodes, including lock box, service bureaus, etc.)	

Identify and list software products and data backups required from off-site storage.

SUPPLIES

Identify and list supplies, including pre-printed forms, computer paper, etc, required at the recovery site. Include prospective suppliers with overnight delivery services.

Attach additional sheets as needed.

IMPORTANT! Consult the User Recovery Center Records Requirements Checklist for a list of records and data that must be secured from storage and delivered to the User Recovery Facility.

Figure 10.3 User recovery center requirements checklist. (Continued)

Notification

In addition to site preparation, the user recovery plan must provide for the notification of users and vendors providing key services and products including supplies, workstation equipment, office equipment, telecommunications rerouting services, and mail and overnight delivery services of the new user work facility address. A contact list should be prepared and included in the plan for later integration into an overall notification directory for the disaster recovery planning document.

Transportation

For user recovery strategies envisioning the use of remote user work facilities, some provision needs to be made for transporting users to the recovery center. Planning includes designation of gathering points, identification of charter bus or plane services, and provisioning for tickets, travel expenses, and so forth.

Shift Schedules

If system and network services are to be available on a restricted basis, it may be necessary for users to perform work in shifts. Tentative shift schedules should be articulated to determine whether system resources are adequate during the times that work must be accomplished.

Food and Shelter

If users will be operating from remote recovery facilities, provisions need to be made for food and shelter of workers while they are not working. Planners need to determine what types of accommodations are available to house workers in proximity to the recovery center.

Food service companies and caterers may need to be identified in advance. Restaurants in the vicinity of the site should be identified and their locations designated on a map for distribution to workers when they arrive at the site.

Food and lodging expenses should be estimated and provisions made to ensure that these expenses can be met by the company during the emergency.

DOCUMENT STRATEGY AND ESTIMATE COSTS

The previous research should be gathered and the total solution documented for presentation to management. Cost estimates for all aspects of the strategy should be provided in the plan.

It is sometimes beneficial to contrast the costs for the preferred user recovery strategy with the potential costs for taking a laissez-faire approach. In addition to the cost of delays in recovery, indicate the difficulties and costs that may be associated with site preparation and vendor negotiation in the midst of an emergency. Even a casual review of the experiences of companies impacted by the World Trade Center bombing or Hurricane Andrew in south Florida that failed to provide a user recovery plan will underscore the consequences of the laissez-faire approach.

SUMMARY

End user recovery planning is often overlooked in writings on disaster recovery planning, but it is extremely important. End user departments now represent significant information repositories in corporations fielding distributed computing platforms. Even in organizations where end users interact with centralized systems, they make all of the systems purposeful. If the recovery of systems is mission critical, so by extension is the recovery of the end user.

Checklist for Chapter

1. Identify business work units that will need to be recovered following a disaster. For each, identify system, network, off-site storage, and recovery personnel requirements.

2. Examine backroom operations strategies. Select appropriate option and develop costs.

3. Determine necessary steps to prepare a backup site for use.

4. Prepare a notification contact list for each end-user department.

5. Document strategy and prepare for presentation to senior management.

Presentation of the Plan Overview

Following the assessment of risks and the formulation of strategies to eliminate avoidable disaster potentials and to minimize the impact of disasters that cannot be eliminated, it is time for the planning coordinator to present findings to management. Senior management approval will be required to realize the disaster avoidance systems and risk mitigation strategies that have been identified and to create a disaster recovery capability for the company.

This activity will include the presentation of an implementation plan, an implementation budget, and an implementation schedule. In short, the disaster recovery project will spawn a series of subprojects that will establish collectively the company's ability to respond to an emergency.

PRESENTATION OF FINDINGS TO SENIOR MANAGEMENT

In Chapter 3, the disaster recovery planner presented a management awareness briefing to solicit support for a formal risk analysis and contingency planning project. In that case, the planner communicated the need for planning as a function of good corporate stewardship and in the context of audit requirements.

Having received the approval of management, the initial phases of disaster recovery planning—risk analysis, disaster prevention systems specification, off-site storage planning, systems recovery planning, network recovery planning, user recovery planning, and insurance planning—were undertaken. The results of these efforts are a collection of findings and plans that require another visit to corporate management for presentation. Having identified what needs to be done to prepare the company to respond to a crisis, the planner must now solicit management approval to realize this capability.

The same principles apply to the presentation of findings from the initial stages of disaster recovery planning as applied to the original awareness briefing. Specifically, the presentation must:

- **Reflect advance planning**—An agenda, visual aids, handouts for meeting participants, and the reservation of needed meeting rooms, presentation equipment, and so forth all communicate to management that this formal presentation is important and well-considered.
- **Be brief and concise**—Planners should remind management that the briefing is in response to their mandate to undertake analysis and planning. Technological discussions should be minimized and business-related findings should be emphasized. A well-rehearsed presentation should not require more than 10 to 20 minutes.
- **Encourage question-and-answer**—The planner should engage the questions of management calmly and authoritatively. It is likely that the planner knows more than the managers about the recovery requirements of the company, having considered them in far greater detail. However, the planner should avoid having discussions of disaster recovery implementation shift into theoretical debates about the merit of corporate data processing directions, outsourcing costs and benefits, and so forth. The plan that is being developed should be presented as a requirement to deal with current exposures and a mechanism for responding to new changes and directions in the company.
- **Embody a positive attitude**—As stated in Chapter 3, the goal of the management presentation is to advise, not to sell. Planners must be prepared to defend their recommendations, but will need to rely on management to make the decision whether to move on to implementation.

The specific content of the presentation will vary from one company to the next, but the presentation typically begins with the summary of activities undertaken according to the original management mandate. Following are the key agenda points for review.

Exposure Analysis/Risk Analysis

Present the methods used to perform risk analysis. Identify the key findings of the analysis and the ramifications of these findings for corporate survivability. Avoid debates on the quantitative methods used to arrive at findings. Invite anyone who objects to the validity of your findings to review the data collected and to perform his/her own analysis. The bottom line approach is best. If a disastrous interruption occurs within the company in its current state of nonpreparedness, the danger is great that no recovery effort will succeed.

Disaster Avoidance Capabilities and Costs

Identify the method used to identify exposures that can be eliminated or minimized through disaster avoidance systems. Identify the method used to evaluate competitive bids to provide the necessary exposure reduction capabilities. Strongly recommend the adoption of disaster avoidance systems as a direct reduction of risks.

This presentation should include off-site storage recommendations. Identify what discoveries were made about unprotected data and the strategy settled upon for mitigating this exposure. The bottom line is that off-site storage of critical data is paramount to any recovery effort.

System Recovery Strategies and Costs

Without belaboring technology issues, identify the key applications supporting the business and their associated platforms. Briefly outline the strategy that has been developed to cope with the possibility of an unplanned interruption of these information services. More than one strategy may need to be fully articulated to position the costs and benefits of the selected strategy into a better context for discussion and adoption.

Network Recovery Strategies and Costs

Telecommunications and data communications networks should be identified and briefly described. Avoid technological discussions, but position the selected strategy for recovering networks in a context that will demonstrate its cost-benefit advantages.

User Recovery Strategies and Costs

Provide a scenario that will demonstrate the need for user recovery planning. Describe the impact of any plan that embodies system and network recovery, but not user recovery. Identify the strategy selected for recovering user work areas and the costs for its implementation.

Insurance Coverages and Costs

Review current coverages by insurance carriers. Explain the advantages of disaster recovery planning for insurance cost reduction. Promote this advantage as an offset to plan costs.

PRESENTATION OF IMPLEMENTATION PLAN

The presentation of findings from the first phases of the planning project is a preamble to the real purpose of the management briefing. Namely, the planner is seeking approval and funding to implement the project plan and to realize a testable disaster recovery capability for the company.

An implementation plan must be presented at this point, identifying what is needed to complete the work that management has initiated. Deliverables inherent in this plan are identified in Figure 11.1. Specific tasks are discussed in the following.

Implementation Tasks

A high-level listing of action items or tasks to be performed in the implementation of the project plan needs to be presented in brief, easily understood terms. These tasks may include the following.

DISASTER AVOIDANCE SYSTEMS

Coordinate and install hazard detection and suppression systems

Train personnel in their use

OFF-SITE STORAGE PROGRAM

Contract for off-site storage services

Articulate and implement program of regular backup and storage

Monitor program participation

DISASTER RECOVERY PLAN

Obtain service contracts

Document plan objectives and procedures

Conduct plan audit

TEAM SELECTION & TRAINING PLAN

Identify disaster recovery team requirements

Staff teams

Develop and schedule training courses

TESTING PLAN AND TEST SCHEDULE

Develop objectives for plan test

Select testing strategy and schedule testing

MAINTENANCE PLAN

Develop change management strategy for plan maintenance

Figure 11.1 Implementation plan deliverables.

Purchase of Disaster Avoidance Capabilities

Identified disaster avoidance systems need to be purchased from vendors. Invite management to review the competitive bids solicited from qualified vendors, but encourage them to approve an approach quickly to reduce the immediate risk confronting normal business operations from fire, flooding, equipment damage due to environmental contamination, and vandalism or theft.

Install Disaster Avoidance Systems

Set forth a plan for installing the disaster avoidance systems that will minimize disruption of normal work. Include subtasks such as the training of key personnel in system use.

Contract for Off-Site Storage

Present the selected off-site storage vendor contract for signature. Emphasize that the implementation of an off-site storage program represents an excellent cost-to-risk reduction ratio. Emphasize that systems and networks can be reconstructed readily, but critical corporate data can be lost forever if not backed up and removed off-site.

Formalize and Articulate Off-Site Storage Schedule

With an off-site storage vendor selected, explain why management support will be required to ensure the best possible use of the program. Emphasize that business unit managers and corporate employees will require a mandate from management in some cases to change their former practices and to participate in the new program. Explain how the recommended schedule will be noninvasive and will not interfere with productivity.

Develop Procedures for Recovering Critical Systems

That a strategy has been developed to provide the best possible system recovery capability in place at the lowest cost does not mean that procedures have been developed for the implementation of the strategy when needed. Time will be required to formalize the procedures and to test them.

Contract with Systems Recovery Vendors

Assuming that a hot or cold site strategy is the strategy recommended for adoption within the company, present the contract to management for review and signature at this time. Indicate that a meeting or on-site visit can be conducted with the vendor if management desires. Indicate the timeframe for purchasing the service at the proposed price.

Develop Procedures for Recovering Networks

As with procedures development for systems recovery, time will also be required to formalize network recovery procedures. Identify what will be required to complete this task and the estimated time to completion.

Contract with Network Recovery Vendors

If on-demand rerouting services, alternative links, or other vended solutions are part of your network recovery plan, have contracts ready to present to management at this time. Indicate whether proposed pricing is subject to expiration.

Develop Procedures for User Recovery

Identify requirements in time and resources to develop user recovery procedures. Indicate that additional requirements for management pronouncements or policy-making will be required to formalize the user recovery capability.

Contract with Vendors for User Recovery-related Services

Identify any vended services that will be used in connection with the user recovery plan and the status of agreements for these services. Submit any existing proposals to management for review and funding, explaining briefly how they will be used in the context of recovery.

Develop Emergency Decision-making Flowchart

Explain that an emergency decision-making flowchart will provide an important management tool for visualizing and coordinating the recovery effort. This flowchart will be prepared and management will be fully briefed in its use.

Develop a Notification Tree

Explain the purpose of a notification tree and describe how key personnel will be notified in the onset and aftermath of a disaster. Development of the notification tree will be an important part of the plan.

Designate and Staff Recovery Teams

Indicate that some personnel will need to be tapped and trained to perform specific functions for system, network, and user recovery. Recommend a strategy for accomplishing this team selection and training function. Solicit management input on performing this task with minimum disruption of daily productive work. Ask whether an incentive program can be implemented to encourage participation.

Train Recovery Teams

Describe team training and outline a corporate awareness program that you would like to see implemented. Solicit management feedback on both.

Test Plan and Revise

Stress that any plan implemented will need to be tested. Reassure management that testing need not require the interruption of business, but that even paper testing will require some time from recovery teams.

Develop Maintenance and Testing Schedule for Disaster Recovery Plan

Explain that a schedule will be developed for ongoing plan maintenance and testing. This is a low-cost maintenance item that will be vital to the plan's acceptance by auditors and other interested parties. Explain that testing will also provide useful information about the evolution and change of business functions and their associated recovery requirements.

Implementation Budget and Schedule

Prepare a written budget that incorporates the costs associated with the tasks set forth previously. The budget need not be presented in the meeting, though some totals may be required to respond to management questions. Additionally, a schedule for task completion and deliverables production should be provided in writing.

Budgeted Expenses

The final phases of the disaster recovery planning project will be more expensive than the preceding phases. This is because specific products and services will be purchased and the number of personnel involved (in training and testing, for example) will grow in number. Budget items may include the following.

Management

The salaries of key project personnel continue to be a cost item in the final implementation stages of the project. Additional expense may accrue to publication-related activities, such as writing and editing the plan as the planning document is articulated more formally.

Technical Support

Direct technical support may be required to review plan provisions or to coordinate installation of plan-related systems, such as fire detection and suppression systems or network monitoring systems.

Clerical Support

Clerical costs may increase as documents are printed, collated, and bound for distribution to senior management, auditors, team members, work unit managers, and so forth.

Aggregate Vendor Costs

Aggregate vendor costs will include the costs for all disaster avoidance systems, an off-site storage program, hot or cold site facilities agreements, network-related services, and user recovery services.

Travel

Travel expenses may accrue to tests performed with the hot site vendor, travel to professional conferences, and so forth. An additional travel-related expense may be incurred if senior managers wish to travel to vendor facilities for on-site inspections prior to contract signing.

Supplies and Materials

These costs will increase as the plan is developed and published. An appropriate budget amount will need to be estimated.

Schedule of Deliverables

A project schedule should be articulated to include all deliverables from the implementation phases of the project. Deliverables serve as project milestones and can be used by management to ensure that the disaster recovery planning project is well-managed and performing useful, meaningful work. Deliverables may include the following.

Disaster Avoidance Capabilities

A schedule for the implementation of each disaster avoidance system, as well as the initiation of the off-site storage program, should be provided. Vendors are useful in assisting with the breakdown of system installation tasks and the estimation of their timeframe for completion.

Disaster Recovery Plan Document

The disaster recovery plan is a complex, yet simple, document. Since most data for the plan has been assembled, a reasonable estimate can be made regarding its publication date.

Team Selection and Training Plan

Team identification, selection criteria, and training requirements should be fairly well-known at this point. A plan will need to be articulated that includes this information. This plan will become a deliverable for the project.

Testing Plan and Test Schedule

A test plan, identifying methods, frequency, and schedules for plan testing, should be created. This deliverable should be included in the project schedule.

Maintenance Plan

A change management strategy will be formulated and documented and will become a deliverable in the final stages of the planning project.

PRESENT COST ESTIMATES FOR MAINTENANCE

The completion and delivery of the previous deliverables constitutes the implementation of the corporate disaster recovery capability. However, the management presentation may also need to include cost estimates for maintenance of the capability over the long haul. Cost items may include the following.

Coordinator Salary

Whether full-, or part-time, there will be an ongoing need for someone to coordinate the maintenance and ensure the currency of the company disaster recovery plan. That person will have a salary.

Vendor Contract Renewal

Trend lines should be reviewed to determine how the costs of vendor-supplied products and services have risen or fallen over the past decade. An estimate of costs over a multiyear period should be made.

Maintenance Contracts on Redundant/Plan-Specific Hardware and Software

Hardware and software cost projections are notoriously inaccurate. However, data-center maintenance agreements can be referenced to obtain an estimate of maintenance contract price performance. This data can then be related to redundant/plan-specific hardware and software maintenance costs.

Training Budget

It may be possible to turn training requirements over to a corporate training department in time. If it is, the training department will be able to provide a cost estimate for training maintenance. If training will be conducted by planning personnel, the budget should be estimated based on salary and resource requirements.

Testing Budget

Different types of testing cost different amounts of money. Predict two tests with the hot site per year and ask the vendor what other customers find their costs to be. Use the vendor's estimate as a starting point and embellish it as needed with your own company's testing requirements.

Insurance Premiums

Determine whether any insurance costs will be reduced once the plan is in place and security/disaster avoidance systems are installed. Include reduced insurance costs in your budget for maintenance.

SUMMARY

Implementation of the disaster recovery planning project, like the initiation of the project, can only successfully occur with the backing of senior management. Planners are well advised to present plans professionally, explain rationales and alternatives carefully, and portray sensitivity to business requirements at all times. Disaster recovery projects cannot be sold to anyone who does not intuitively perceive their value.

Checklist for Chapter

1. Prepare an agenda for presentation to senior management of the following:

    - Risk analysis methods and results
    - Disaster prevention systems analysis and progress
    - Off-site storage planning progress
    - Systems recovery strategy and costs
    - Network recovery strategy and costs
    - End-user recovery strategy and costs

2. Prepare implementation plan. Identify tasks to be performed and costs.

3. Prepare maintenance plan. Identify maintenance tasks, schedule, and costs.

4. Prepare presentation materials and rehearse presentation.

5. Schedule management briefing. Make presentation.

Plan Development

Having obtained management approval to proceed with implementation of the strategies formulated to eliminate or minimize risks confronting the company, it remains to develop the procedures for specific recovery activities and an emergency decision-making flowchart to coordinate them.

This chapter sets forth one strategy for designing and developing a plan document. Emphasis is placed on the ease of use and ease of maintenance of the document. Despite the claims of some vendors of "canned plans"—microcomputer-based disaster recovery planning tools—all that is needed is a word processor, a graphics software tool, and the data collected in the initial phases of the planning project. Once written, the plan can be maintained with the same tools and data obtained from plan testing and change management forms.

STRUCTURE OF THE PLAN DOCUMENT

There are as many approaches to developing plan documents as there are planners. Each document is a carefully crafted report of strategies, broken into tasks and procedures, and an emergency decision-making flowchart that provides the associative glue. Some planners prefer to place the emergency management flowchart at the front of the document and all subordinate procedures as follow-on material. They claim this facilitates the use of the document in an emergency situation.

In the structure of the plan provided in this book, an index is provided prior to the flowchart and procedures. This is to facilitate maintenance of the plan and its use by auditors. It is expected that these two uses will be far more common than the use of the plan in response to a disaster. If an actual disaster occurs, it is a simple matter to turn to the emergency management section of the plan and proceed with the documented strategy at that point. Figure 12.1 provides the structure for a plan document used in the service of many companies.

Both approaches are well-taken, however, and there is no right or wrong way to develop a plan. The only incorrect way is not to plan at all.

Considerations for Ease of Maintenance

Ease of maintenance is extremely important in disaster recovery plan development. The documented plan is subject to regular changes as the business it is designed to protect evolves and changes. In general, variable information should be grouped together. Names, vendor contacts, phone and address information, and other items need to be placed in a single location of the document where they can be updated simply.

ONE STRUCTURE FOR A DISASTER RECOVERY PLAN

INDEX TO THE PLAN

EMERGENCY RESPONSE PROCEDURES
NOTIFICATION DIRECTORY AND EMERGENCY DECISION-MAKING FLOWCHART

SYSTEM RECOVERY PROCEDURES

NETWORK RECOVERY PROCEDURES

USER RECOVERY PROCEDURES

APPENDIX A
SUMMARY OF RISK ANALYSIS AND OVERVIEW OF AVOIDANCE AND RECOVERY STRATEGIES

APPENDIX B
OFF-SITE STORAGE PLAN AND SCHEDULE

APPENDIX C
TESTING, TRAINING, AND PLAN MAINTENANCE HISTORY

Figure 12.1 One structure for a disaster recovery plan.

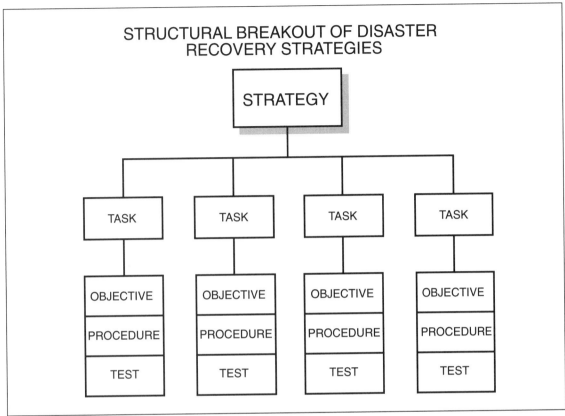

Figure 12.2 Structural breakout of disaster recovery strategies.

Considerations for Ease of Use

Just as variable information should be kept together for reasons of ease of maintenance, so all other information should be grouped logically and intelligibly for ease of use. In the plan structure depicted here, the plan document is divided logically into procedures for system recovery, procedures for network recovery, and procedures for user recovery.

All procedures are derived from tasks. As part of the formulation of each strategy, a collection of tasks were identified. Each task, in turn, incorporates an objective, procedure, and test. Figure 12.2 depicts this structural breakdown of recovery strategies.

For organizational purposes, a numbering technique may be used to identify tasks and their associated procedures. This will aid in the depiction of task interrelationships in an emergency decision-making flowchart. The systems recovery strategy, for example, may be assigned the numerical value of 1.0. As part of this strategy, there may be a task in which the hot site vendor is notified that the company is declaring an emergency. If this is the first task, it may be labelled 1.1. Steps of the procedure associated with the disaster declaration task may receive the values 1.1.A, 1.1.B, 1.1.C., and so forth.

TASK OBJECTIVES/PROCEDURES WORKSHEET

STRATEGY: _____

TASK: _____

OBJECTIVE: _____
_____

PREREQUISITES/ _____
DEPENDENCIES: _____
_____

PROCEDURE TO COMPLETE TASK AND REALIZE OBJECTIVE

STEP	DESCRIPTION	REQUIREMENTS	DEPENDENCIES

TASK COMPLETE? (Circle One)   YES   NO
If NO, please attach additional pages to provide complete procedure for accomplishing the task.

PAGE ____ **OF** ____

Figure 12.3 Task objectives/procedures worksheet.

Each strategy section of the plan should provide a brief summary of the objectives of the strategy and a list of all associated tasks and their objectives. This should be readily accomplished using the analysis conducted when formulating the strategy.

All that remains to complete a strategy section is to complete the task procedures. Figure 12.3 provides a worksheet for documenting the procedural tasks associated with each task. These sheets may be added in sequence behind the strategy overview and task list.

Depending on the complexity of the plan, other logical groupings of tasks may be required. A subsection may be required for branch systems recovery, for example, or for restoring a particularly complex system. The important point is to group tasks logically and in such a way that they can be used expediently in an emergency, for testing, and for training.

Other parts of the plan may include appendices summarizing the risk analysis conducted by the project team, the details of the off-site storage plan, and historical notes on plan testing, training, and maintenance. The last appendix is particularly useful in audits.

EMERGENCY DECISION-MAKING FLOWCHART

It may be noticed in Figure 12.3, that the form provides a location for noting dependencies or prerequisites for the task at hand. In an actual recovery multiple teams work multiple tasks concurrently. For one task to be completed, another task—possibly in another strategy area—must be completed first. These prerequisite tasks are noted on the task form.

Interdependencies may be identified readily using a PERT or GANT chart to depict recovery tasks. A planner can endeavor to construct one using any drawing tool or any common project management software package may be used. A PERT or GANT chart for a strategy may be included in each strategy section of the plan.

Combining all strategy-specific PERT or GANT charts into a single chart results in an emergency decision-making flowchart. This chart may be simplified for nontechnical use or it may be constructed to the finest level of detail to provide technical managers with an exact accounting of the sequences and interrelationships of all tasks.

Figure 12.4 provides an example of an extremely high-level GANT chart depicting the disaster recovery process. As indicated in the chart, the disaster recovery process may be conceived as a three-phase process. The phases are Reaction, Recovery, and Transition.

Figure 12.5 depicts a part of an emergency decision-making flowchart that might be developed from GANT or PERT charts.

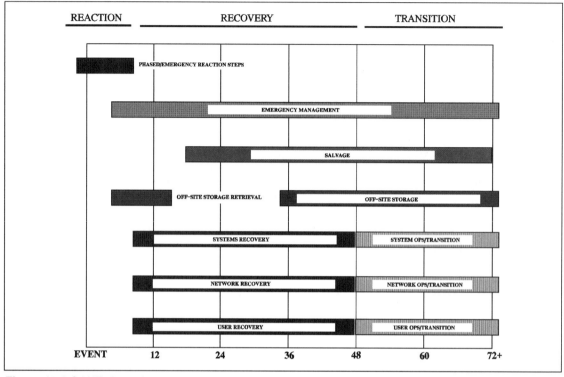

Figure 12.4 GANT chart of a disaster recovery process.

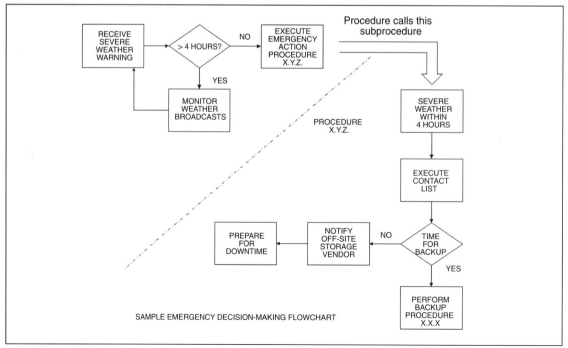

Figure 12.5 Emergency decision-making flowchart.

Reaction Phase

In the Reaction Phase, the company provides a response to the onset of a disaster. Tasks may include hot site notification, emergency management team notification, site damage and outage assessments, off-site storage retrieval, and so forth.

A distinction is made between a phased reaction and an emergency reaction. Phased reaction tasks are a series of steps taken methodically in advance of a disaster. For example, confronted with a hurricane emergency, the company may have several hours to perform last minute backups, to power down and cover hardware, and to perform other procedures intended to reduce the disastrous consequences of the actual storm.

Other emergencies, such as sudden fires or a roof collapse, may occur without advanced warning. The steps taken in response to the emergency may be far less proactive. However, tested procedures will help to ensure that the reaction to the disaster is no less methodical.

Recovery Phase

The Recovery Phase entails all procedures for recovering mission-critical systems, networks, and user operations within the timeframe specified in the plan. Companies can operate from hot sites, alternative networks, and backroom operations centers for a considerable period of time while salvage operations are undertaken or new facilities are located and equipped.

Transition Phase

The Transition Phase is rarely documented in disaster recovery plans as its parameters are not known until a disaster has occurred. In some cases, companies utilizing hot sites will be required to relocate to cold or shell sites provided by the same vendor after a specific period of time. This may entail a set of tasks (i.e., ordering replacement hardware, installing hardware, redirecting networks) that can be documented in the plan.

To the extent that transition activities are known, they should be documented to reduce the unknowns confronting management following a disaster.

TEAM SELECTION

Figure 12.4 suggests that the disaster recovery process will entail concurrent tasks performed by individuals or teams working in concert. Nearly every modern disaster recovery plan utilizes a team approach for accomplishing critical tasks. The designation of these teams varies from one plan to the next, depending on unique recovery requirements.

In general, there will be an emergency management team responsible for the command, control, and communication aspects of the recovery effort. Members of this team may include a designated disaster recovery coordinator, senior management representatives, a financial executive, and the company's press relations agent.

The emergency management team will set up an emergency management center that will be responsible for receiving field reports and coordinating the efforts of recovery teams, vendors, insurance companies, and so forth. This management effort is guided by the emergency decision-making flowchart.

Typically, the emergency management team will also set up an emergency communications center. This will be the central point for all media management and for fielding the inquiries of investors, customers, and key clients.

In addition to emergency management, other teams may include a wide-area network restoration team, a mainframe systems recovery team, a LAN restoration team, a user-center preparation team, a salvage team, an off-site storage team, and so forth. Each team is defined by the tasks it must perform. It has a mission, a timeframe for completing work, and numbers among its members personnel with skills and knowledge to perform the assigned tasks correctly.

All teams should have a team leader and a designated alternate team leader. These persons are responsible for participating in all plan tests and for developing an appreciation for all of the tasks assigned the team and their interrelationships with the tasks performed by other teams.

In an emergency, the team leader or alternate will be responsible for contacting other team members and for initiating work as soon as the emergency management team gives the word. As part of plan maintenance, team leaders will ensure that procedures and team member positions remain current with the changing business and with staff turnover.

Figure 12.6 provides a form for recording team membership. This form should be part of the emergency management section of the plan and should be reviewed for its currency in each iteration of the plan document.

Criteria for Selection

The designation of team leaders, alternates, and members is not arbitrary. Most recovery tasks require personnel with an intimate knowledge of the technologies or business procedures that are to be restored. Team membership generally is decided on the basis of whether certain prerequisite skills and knowledge are present in the team candidates.

At the same time, it should be noted that no criteria other than skills and knowledge is immediately available to aid in selection. Team leaders may be male or female. They may be single or married. These factors do not impair or improve their capabilities in a crisis.

Membership may be guided by practical considerations, however. If a prospective member lives within a certain proximity to the company headquarters, that candidate may be preferable (or less preferable) to one who lives further away, or to one who must cross bridges and so forth, to get to the recovery site. Member candidates who have fears of air travel may not be good choices to fly to a hot site in an emergency.

Myriad considerations may enter into team selection. The ultimate consideration, however, is whether the candidate possesses the knowledge and skills—and perhaps the creativity—to accomplish the tasks identified for the team.

TEAM DEFINITION WORKSHEET

STRATEGY: _____
TASKS: _____
TEAM NAME: _____
TEAM PURPOSE: _____
_____
SKILLS/KNOWLEDGE: _____
_____
_____

TEAM LEADER

NAME: _____

OFFICE PHONE: _____
HOME PHONE: _____
ALTERNATE PHONE: _____
PAGER/CELL PHONE: _____

ALTERNATE TEAM LEADER

NAME: _____

OFFICE PHONE: _____
HOME PHONE: _____
ALTERNATE PHONE: _____
PAGER/CELL PHONE: _____

NAME: _____

OFFICE PHONE: _____
HOME PHONE: _____
ALTERNATE PHONE: _____
PAGER/CELL PHONE: _____

NAME: _____

OFFICE PHONE: _____
HOME PHONE: _____
ALTERNATE PHONE: _____
PAGER/CELL PHONE: _____

Page 1 of ___

Figure 12.6 Team definition worksheet.

NAME: ______________________

OFFICE PHONE: ______________________
HOME PHONE: ______________________
ALTERNATE PHONE: ______________________
PAGER/CELL PHONE: ______________________

NAME: ______________________

OFFICE PHONE: ______________________
HOME PHONE: ______________________
ALTERNATE PHONE: ______________________
PAGER/CELL PHONE: ______________________

NAME: ______________________

OFFICE PHONE: ______________________
HOME PHONE: ______________________
ALTERNATE PHONE: ______________________
PAGER/CELL PHONE: ______________________

NAME: ______________________

OFFICE PHONE: ______________________
HOME PHONE: ______________________
ALTERNATE PHONE: ______________________
PAGER/CELL PHONE: ______________________

NAME: ______________________

OFFICE PHONE: ______________________
HOME PHONE: ______________________
ALTERNATE PHONE: ______________________
PAGER/CELL PHONE: ______________________

Page 2 of ___ **(Attach additional pages as needed)**

Figure 12.6 Team definition worksheet. (Continued)

Role of Human Resources

Human Resources can play a role in keeping the team membership rosters up-to-date. When an employee resigns or is terminated, Human Resources should conduct an exit interview or maintain a personnel file that will identify whether the employee was a member of the company disaster recovery team. Notification should be directed to the Disaster Recovery Coordinator to ensure that the team position is filled and that the new team member is trained.

NOTIFICATION TREES

As previously indicated, team definition forms should be included in the emergency management section of the plan document. This will facilitate the notification of team leaders or alternates in an emergency. These persons will then be responsible for notifying all team members.

The emergency management section of the disaster recovery plan should also contain other notification directories. These include an Emergency Management Notification Directory (Figure 12.7), a Departmental Notification Directory (Figure 12.8), and a Vendor Notification Directory (Figure 12.9).

Specific responsibilities should be assigned in the plan for contacting various persons on the list in order to activate all personnel required to recover business operations. These lists should be reviewed and maintained at routine intervals as part of the change management procedure.

In some plans, a graphical depiction is made of notification responsibilities. Person A notifies persons B, C, and D, who in turn notify persons E, F, and G. The result is a tree-like chart representing a cascade of notifications. The problem with this approach is that a breakdown in communications might occur if one contact misunderstands or misquotes to the next three, and they in turn misquote to the next tree branch, and so forth.

To avoid this problem, it may be useful to establish a simple contact procedure that notifies every person that an emergency exists. They may then call a designated hot line (many disaster recovery vendors provide them) for additional information and orders. This single number approach also offers the advantage of involving personnel quickly, even if they were unavailable during the initial round of contacts.

EMERGENCY MANAGEMENT NOTIFICATION DIRECTORY

SENIOR CRISIS MANAGER

NAME: _____

OFFICE PHONE: _____
HOME PHONE: _____
ALTERNATE PHONE: _____
PAGER/CELL PHONE: _____

DESIGNATED CRISIS MGR ALTERNATE

NAME: _____

OFFICE PHONE: _____
HOME PHONE: _____
ALTERNATE PHONE: _____
PAGER/CELL PHONE: _____

TECHNICAL OPERATIONS MGR

NAME: _____

OFFICE PHONE: _____
HOME PHONE: _____
ALTERNATE PHONE: _____
PAGER/CELL PHONE: _____

COMMUNICATIONS MANAGER

NAME: _____

OFFICE PHONE: _____
HOME PHONE: _____
ALTERNATE PHONE: _____
PAGER/CELL PHONE: _____

RECOVERY FINANCE COORDINATOR

NAME: _____

OFFICE PHONE: _____
HOME PHONE: _____
ALTERNATE PHONE: _____
PAGER/CELL PHONE: _____

Figure 12.7 Emergency management notification directory.

NAME: _____

OFFICE PHONE: _____
HOME PHONE: _____
ALTERNATE PHONE: _____
PAGER/CELL PHONE: _____

NAME: _____

OFFICE PHONE: _____
HOME PHONE: _____
ALTERNATE PHONE: _____
PAGER/CELL PHONE: _____

NAME: _____

OFFICE PHONE: _____
HOME PHONE: _____
ALTERNATE PHONE: _____
PAGER/CELL PHONE: _____

NAME: _____

OFFICE PHONE: _____
HOME PHONE: _____
ALTERNATE PHONE: _____
PAGER/CELL PHONE: _____

NAME: _____

OFFICE PHONE: _____
HOME PHONE: _____
ALTERNATE PHONE: _____
PAGER/CELL PHONE: _____

(Attach additional pages as needed)

Figure 12.7 Emergency management notification directory. (Continued)

DEPARTMENTAL NOTIFICATION DIRECTORY

WORK AREA: _____

NUMBER OF PERSONNEL: _____
NUMBER OF RECOVERY
PERSONNEL: _____

PRIMARY CONTACT: _____
SECONDARY CONTACT: _____
NOTIFIED BY: _____
PER PLAN PROCEDURE: _____

NAME & TITLE   **PHONE & EXT**

_____   _____
_____   _____
_____   _____
_____   _____
_____   _____
_____   _____
_____   _____
_____   _____
_____   _____
_____   _____
_____   _____
_____   _____
_____   _____
_____   _____
_____   _____
_____   _____
_____   _____
_____   _____

Attach additional pages as needed

Figure 12.8 Departmental notification directory.

VENDOR NOTIFICATION DIRECTORY

HOT SITE VENDOR

VENDOR: _____ **PHONE & EXT**

PRIMARY CONTACT: _____ _____
SECONDARY CONTACT: _____ _____
NOTIFIED BY: _____
PER PLAN PROCEDURE: _____

BACKROOM OPS CENTER

VENDOR: _____ **PHONE & EXT**

PRIMARY CONTACT: _____ _____
SECONDARY CONTACT: _____ _____
NOTIFIED BY: _____
PER PLAN PROCEDURE: _____

NETWORK SERVICES

VENDOR: _____ **PHONE & EXT**

PRIMARY CONTACT: _____ _____
SECONDARY CONTACT: _____ _____
NOTIFIED BY: _____
PER PLAN PROCEDURE: _____

OFF-SITE STORAGE

VENDOR: _____ **PHONE & EXT**

PRIMARY CONTACT: _____ _____
SECONDARY CONTACT: _____ _____
NOTIFIED BY: _____
PER PLAN PROCEDURE: _____

HARDWARE VENDOR

VENDOR: _____ **PHONE & EXT**

PRIMARY CONTACT: _____ _____
SECONDARY CONTACT: _____ _____
NOTIFIED BY: _____
PER PLAN PROCEDURE: _____

Page 1 of ___

Figure 12.9 Vendor notification directory.

VENDOR: _____   **PHONE & EXT**

PRIMARY CONTACT: _____   _____
SECONDARY CONTACT: _____   _____
NOTIFIED BY: _____
PER PLAN PROCEDURE: _____

VENDOR: _____   **PHONE & EXT**

PRIMARY CONTACT: _____   _____
SECONDARY CONTACT: _____   _____
NOTIFIED BY: _____
PER PLAN PROCEDURE: _____

VENDOR: _____   **PHONE & EXT**

PRIMARY CONTACT: _____   _____
SECONDARY CONTACT: _____   _____
NOTIFIED BY: _____
PER PLAN PROCEDURE: _____

VENDOR: _____   **PHONE & EXT**

PRIMARY CONTACT: _____   _____
SECONDARY CONTACT: _____   _____
NOTIFIED BY: _____
PER PLAN PROCEDURE: _____

VENDOR: _____   **PHONE & EXT**

PRIMARY CONTACT: _____   _____
SECONDARY CONTACT: _____   _____
NOTIFIED BY: _____
PER PLAN PROCEDURE: _____

(Attach additional pages as needed)

Figure 12.9 Vendor notification directory. (Continued)

AN EXAMPLE FROM A DISASTER RECOVERY PLAN

The guidance provided in this chapter has been deliberately nondirective. This is intentional. To provide a specific format for the disaster recovery plan would be the same as saying that there is only one way to write a plan. In fact, plans should be unique and reflect the particular requirements of a company. For those who need additional examples, a sample table of contents for a real-world disaster recovery plan is provided in Figure 12.10.

ACME CORPORATION
DISASTER RECOVERY PLAN
TABLE OF CONTENTS

0.0   Emergency Response
    0.1   IF this is an ADVANCED WARNING situatio, GOTO Flow Diagram X
    0.2   IF this is a NO WARNING situation, GOTO Flow Diagram Y

1.0   Decision-making Flowcharts
    1.1   Master Chart
    1.2   Disaster Affecting Data Center—RESPONSE PHASE
    1.3   Disaster Affecting Headquarters—RESPONSE PHASE
    1.4   Disaster Affecting Branch Offices—RESPONSE PHASE
    1.5   Disaster Affecting Data Center—RECOVERY PHASE
    1.6   Disaster Affecting Headquarters—RECOVERY PHASE
    1.7   Disaster Affecting Branch Offices—RECOVERY PHASE

2.0   System Restoration Strategy and Procedures
    2.1   Data Center Restoration
    2.2   Distributed System Restoration

3.0   Network Restoration Strategy and Procedures
    3.1   Voice Restoration
    3.2   Data Network Restoration

4.0   End User Recovery Strategy and Procedures
    4.1   Evacuation Procedures
    4.2   Site Preparation Procedures
          4.2.1   Communications Center
          4.2.2   Emergency Management Center
          4.2.3   Backroom Operations Center

5.0   Notification Directories
    5.1   Disaster Recovery Personnel
    5.2   Senior Management/Business Unit Management
    5.3   Vendors
    5.4   Emergency Response Organizations and Utilities
    5.5   Key Customers

6.0   Plan Testing and Revision History (Audit Copy Only)
    6.1   Risk Analysis
    6.2   Off-Site Storage Plan
    6.3   Vendor Contracts
    6.4   Change Management
    6.5   Test History

Figure 12.10 Sample table of contents for a disaster recovery plan.

SUMMARY

There is no official or approved structure for an effective disaster recovery plan document, rules of thumb apply: make it easy to use and easy to maintain. Beyond this, models of plans are available in print or on disk from numerous sources which can help planners visualize a useful document. Be aware, however, that automated planning tool formats for disaster recovery plans typically exclude network and end-user recovery procedures. Neglecting these important components will likely produce a document better suited for use as a door stop than a disaster recovery guide.

Checklist for Chapter

1. Develop a structure and table of contents for the disaster recovery planning document. Identify how it will facilitate maintenance and use in an emergency.

2. Develop an emergency decision-making flowchart and map plan tasks to specific chronological events.

3. Develop structure diagrams to depict work required to realize the objectives and tasks of each disaster recovery strategy: systems recovery, network recovery, and end user recovery.

4. Define recovery teams and their requisite skills. Select team leaders and members.

5. Implement a procedure for notification by Human Resources of employee turnover that affects the disaster recovery personnel complement.

13

Disaster Recovery Training

More than one disaster recovery planner has told the story of a consulting firm—often one affiliated with a Big Six accounting firm—that has offered to complete the company's disaster recovery plan to satisfy an audit requirement. Often, the sales pitch provides for a comprehensive disaster recovery plan filling an impressive three-ring binder bearing the logo of the firm and guaranteed to pass muster in the next audit.

The price for the service is steep, but so is the price of not having a plan successfully completed in time for the next audit—a real possibility if the firm's experts do not write the plan using their preexisting knowledge of company systems, networks, and vulnerabilities. The price can be reduced, the marketeer offers in conspiratorial tones, if the testing and training phases are omitted.

The irony is glaring. Testing the plan is absolutely vital for ensuring its viability. Planning is an iterative process with periodic tests providing the means to identify omissions. The details of testing and of change management are covered in Chapter 14.

Training is equally vital. The disaster recovery plan is little more than a massive paper weight or an impressive door stop if no one knows how to implement its procedures in the face of an unplanned interruption in business activity. Training imparts a knowledge of the plan to those who must enact it in an emergency. Moreover, training can have a prophylactic benefit when provided in the form of a corporate employee awareness program. Employees, trained to recognize disaster potentials, are the first line of defense in corporate survivability. They can report hazards so steps can be taken to eliminate them.

In this chapter, we will look at the elements of a successful disaster recovery training program. The discussion will encompass both traditional forms of corporate training and training opportunities available outside of the company.

THE DISASTER RECOVERY PROJECT
STAFFING AND TRAINING

MAJOR ACTIVITIES

Identify Recover Teams

Select Team Members

Develop Notification Tree

Develop Emergency Management Flowchart

Train Team Members

Implement Corporate Disaster
Awareness Program

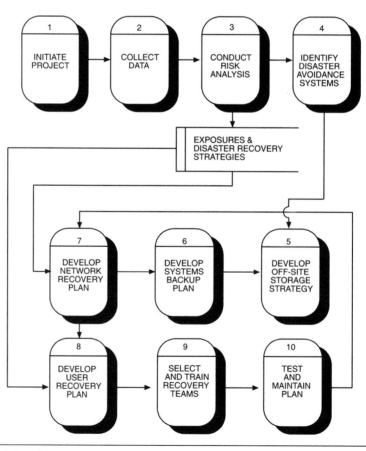

Figure 13.1 The staffing and training phase.

OBJECTIVES OF TRAINING

The broad objective of training is to create a disaster recovery capability within a company by creating a culture of threat and safety awareness among company personnel. In more practical terms, training strives to accomplish two things:

- Familiarize personnel who will be involved in the implementation of the disaster recovery plan with the intricacies of plan procedures and strategies.
- Increase awareness of disaster potentials and safety risks among corporate personnel so that they can participate in preventing avoidable disasters.

Creating training programs to meet these objectives is part of the staffing and training phase of the disaster recovery planning project as depicted in Figure 13.1. The other activities of this phase, which should already have been accomplished by the time that training programs are formulated and conducted, include identifying recovery teams that will be assigned tasks in the actual crisis, selecting the members of these teams, developing a strategem for notifying team members in the face of a disastrous interruption of mission-critical business functions, and developing an emergency management flowchart to coordinate the activities of teams. These activities have been addressed in previous chapters.

Once recovery teams have been identified and staffed—an activity that may take different forms in different organizations—team leaders and/or members should take part in training on plan procedures. The following section covers techniques for developing training programs both for classroom and text presentation.

DEVELOPING THE TRAINING PROGRAM

The training programs described in the following section are developed in accordance with Instructional Systems Design (ISD) methodology. ISD methodology offers a rational model for training program development that is objective-based. In other words, each lesson in the training curriculum is derived from one or more training objectives. Subject matter is then sequenced for presentation to trainees in a manner that will facilitate learning. Finally, trainee feedback and performance observation is used to evaluate training effectiveness and to revise and improve training programs.

The ISD approach maps well to the structured design of the disaster recovery plan itself. Plan procedures are defined by task and objective and organized under strategies for recovering specific business functions or systems. The task of the training program will be to present these procedures so that procedural objectives correlate to training objectives.

DRS OBJECTIVES DEVELOPMENT WORKSHEET

OBJ #	OBJECTIVE	DR TASK #	TASK SUMMARY	TRAINING METHOD/TIME REQUIRED	VALIDATION

PAGE 1 OF _____

Figure 13.2 DRS objectives development worksheet.

Figure 13.2 provides a useful tool for correlating objectives from the disaster recovery plan to training objectives. For each training objective (OBJ#) and description (OBJECTIVE) there is an associated column for recording the disaster recovery plan task number (DR TASK #). The balance of the form is used to record the teaching method that will be used to communicate the contents and procedures for the task and the method that will be used to validate that learning has occurred.

Of course, training in procedures should be done after trainees have received an overview of plan strategies and have been thoroughly familiarized with disaster recovery plan document contents and design. These preliminary lessons will provide a context for understanding the more rarified elements of procedural training and will enable trainees to use the training guide—the disaster recovery plan document—more effectively.

Classroom

The classroom provides an excellent forum for communicating plan tasks and procedures to those who will need to enact them in an emergency. In a sense, it is a microcosm of an actual disaster recovery situation, with recovery team members interacting much as they would in a real emergency, guided by an instructor who assigns tasks as any emergency manager would in a real crisis.

The success of classroom training will have much to do with the preparation of the instructor and the respect accorded to the trainees as adult learners. General guidelines for conducting a successful classroom training program include the following.

- **Develop a lesson plan**—Figure 13.3 provides an example. Lesson plans allow you to organize training methods, reading materials, trainee requirements, and teaching time estimates in a single place for ease of use. Lessons should correlate to actual disaster recovery plan tasks and should be relevant to the specific audience being trained.
- **Create an agenda**—An agenda shows respect for adult learners by identifying what is to be accomplished in a given timeframe. The training agenda should include start and stop times, break times, and key lessons to be learned. It should also include question and answer times, though questions should be entertained throughout the presentation to ensure that a facilitative environment is maintained. Articulate an agenda and stick with it.
- **Coordinate training with managers**—Ensure that the managers of trainees know that the training schedule will not allow for interruptions. To be successful, classroom training must be conducted with a minimum of outside interruptions.
- **Be mindful of trainee knowledge and experience**—Adult learners require respect for their existing knowledge and experience. Observations, storytelling, and questions should be entertained at all times, provided they are on the subject and do not take away from the time available to communicate key training points. Also, be aware that most disaster recovery training is not about the errata of procedures. Most attendees already know how to perform the procedures identified in the plan. Training helps

learners to understand the criticality of their assigned tasks and the inter-relationships between their tasks and the work of other teams in a crisis.

- **Vary the training methods employed**—Classroom training is not lecturing. Where possible, encourage interaction to enhance understanding. Also, use visual, audio, and other teaching methods as much as possible. Studies show that learning occurs more readily and information is retained longer if multiple senses are used in collecting the information in the first place. Entertainment is also key. Inoffensive joke-telling and "war stories" about the experiences of other companies in disaster recovery situations appeal to learners and help to relax them so that learning can more easily occur.
- **Solicit feedback from trainees**—An evaluation form is provided in Figure 13.4 that may be customized to solicit comments. Trainee evaluations can be a useful tool in understanding which teaching approaches worked and which did not. This input can be used to improve subsequent training lessons.

Of course, these are only a few ingredients of a successful classroom training effort. In some companies, a formal organ for corporate training is available and may be used to develop and conduct disaster recovery team training. At a minimum, the company's training professionals may be able to offer advice on constructing a well-developed and delivered training program.

Text-based

In some organizations, classroom training is not available either due to lack of time, facilities, or resources. Text-based training provides an alternative.

Text-based lessons are developed in the same manner as classroom lessons, except that all "teaching" occurs through the reading of a document—typically, sections of the disaster recovery plan. Following the reading, the trainee may be asked to respond to several written questions about what he/she has read or, at a minimum, to sign a form to signify that the training has been taken.

Clearly, text-based instruction will be validated, as will all training, through trainee participation in disaster recovery plan testing. Trainees who understand what is expected of them in a test scenario and who can find and use the procedures set forth in the plan document have realized the goals of training.

DRS LESSON PLAN

LESSON NUMBER: _____

LESSON TITLE: _____

TASK NUMBER: _____

TASK OBJECTIVES: _____

_____

_____

_____

OBJ #	PROCEDURAL STEP/TEACHING POINT	TRAINING METHOD	MATERIALS REQUIRED

DR PLAN REFERENCES: _____

VALIDATION METHODS: _____

_____

LESSON PREPARED BY: **DATE:**

Figure 13.3 DRS lesson plan.

DRS TRAINING EVALUATION FORM

LESSON NUMBER: _____
LESSON TITLE: _____
INSTRUCTOR: _____
DATE: _____

INSTRUCTIONS: Please complete the following evaluation form to aid us in improving the training provided to disaster recovery team members on the corporate disaster recovery plan and your role within it. Your comments are appreciated.

1. Was the training you received relevant to your role in the corporate DR plan?
    Please circle your choice.

    YES   NO

    If you answered NO, please explain your response:

    _____
    _____
    _____

2. Was training conducted professionally? Was the trainer prepared? Were schedules observed?

    YES   NO

    If you answered NO, please explain your response:

    _____
    _____
    _____

3. Would you change anything about the training if you had the opportunity?

    YES   NO

    If you answered YES, please explain your response: Write on the back, if needed.

    _____
    _____
    _____

Thank you for your participation in this survey!

OPTIONAL: *Please sign and date your evaluation form , if desired. You may be contacted for a follow-up interview.*

PARTICIPANT: _____   **DATE:**

Figure 13.4 DRS training evaluation form.

AWARENESS PROGRAMS

In addition to training management, team leaders, and team members to activate the plan and implement it in an emergency, training of all corporate personnel is also important. The safety and well-being of employees is a primary objective of disaster recovery planning. Thus, they should be familiarized in broad strokes with the plans the company has for responding to an emergency.

At a minimum, employees should know what to do if a hazardous situation is observed, how to react to an emergency evacuation, and whom to contact in the event that a disaster impacts the office where they are to work. In some companies, these important training tasks are given to human resources personnel. A new employee orientation day may provide familiarization with emergency procedures as part of its agenda. In other cases, business unit managers hire and train their own employees and the task of increasing employee awareness is theirs.

Whatever the method currently in use, the disaster recovery project team may seek management cooperation in strengthening it. Posters placed in congregating areas such as coffee rooms, lavatories, and so forth, can be used to reinforce awareness. Suggestion boxes, e-mail addresses for anonymous comments, and other vehicles to facilitate communications regarding disaster potentials can be established and introduced to all employees. Corporate senior managers can assist by making policy statements, writing memoranda applauding the work of disaster recovery planners, and by establishing an Emergency Management Awareness Day at the company.

All of these preparations, including classroom and text-based training of DR teams, help to prepare the personnel of the company to cope rationally with the great irrationality of a disaster. This may be the edge that is needed to respond effectively to an emergency and to get the business up and running following a disaster.

OTHER OPPORTUNITIES FOR EDUCATION

In addition to training and corporate awareness activities undertaken internally, there are other opportunities for disaster recovery training. These include professional organizations for contingency planners, data processing and networking user groups, and trade publications focusing on disaster recovery planning issues.

Professional Contingency Planning Organizations

Nearly every metropolitan area of North America is served by some form of contingency planning group. These are typically professional associations that offer informative agenda on a monthly, bi-monthly and/or annual basis related to disaster recovery planning. The largest group is the Association of Contingency Planners (ACP), based in Los Angeles, California, which has chapters in major cities across the United States.

Corporate disaster recovery coordinators should seek to participate in these organizations and to solicit the participation of disaster recovery personnel and senior management. Dues are generally reasonable for membership and members often receive discounts on annual expositions and conferences.

Participation provides the means to network with fellow planners and to learn from the experiences of others in important areas such as vendor performance, recovery strategies, testing strategies, and obtaining senior management cooperation. If an organization is not conveniently co-located to your company, consider starting one yourself.

User Groups

Vendors of disaster recovery products and services often have user groups for their customers that focus on other aspects of disaster recovery. Planners should check with their hot site vendor, disaster recovery planning tool vendor, network backup vendor, and so forth to see what groups they sponsor and how the company can become a member.

Additionally, most large technology vendors (such as IBM, Hewlett Packard, Digital Equipment Corporation, AT&T, etc.) offer special interest groups and user groups for their clients and customers. While the meetings of these groups may not focus on disaster recovery planning exclusively, a little encouragement from one or more users may be all that is required to put disaster recovery planning on the agenda. The result may be an informative presentation by internal experts in DR within the vendor's organization.

Publications

Several publications are available that focus on disaster recovery planning. Books in this field are becoming much more numerous today than they were 10 years ago. Most computer and telecommunications trade journals include disaster recovery planning on their editorial calendars on a semi-frequent basis.

Planners may wish to begin circulating information on disaster recovery planning, culled from these sources, to senior management and to disaster recovery team leaders on a routine basis. Such activity tends to reinforce training and awareness and may stimulate improved functioning of the disaster recovery change management system.

SUMMARY

Shakespeare once wrote, "All things are ready if our minds be so." Training is an important component of disaster preparedness. Personnel who will be involved in implementing recovery strategies directly, and all the rest who work in the company, need to be trained to recognize and respond to disaster potentials and situations.

Formal programs need to be articulated to address this training requirement. They should be developed according to structured, Instructional Systems Design methods. They should be presented in a manner that enhances their efficacy to adult learners. Only then can avoidable disasters be avoided and the impact of unavoidable disasters minimized.

Checklist for Chapter

1. Develop training objectives based on the plan document.

2. Develop lesson plans for training classes.

3. Create an agenda and schedule for class after consultations with disaster recovery personnel managers.

4. Develop training materials and rehearse.

5. Deliver training courses.

6. Solicit feedback and modify lesson plans and presentation materials appropriately.

7. Develop a corporate disaster awareness program. Solicit senior management approval and implement.

8. Identify opportunities for additional education, including conferences, professional associations, and publications. Involve recovery team personnel in training endeavors.

14

Testing Disaster Recovery Plans

PURPOSE AND SCOPE

For disaster recovery plans to provide a capability for corporate recovery, they must be tested. Tests both validate the integrity of documented recovery procedures and strategies and acquaint test participants with their roles in the event of a disaster.

This chapter examines the subject of plan testing and seeks to answer the questions why, when, and how. The module will also examine the roles of vendors, business managers, auditors and recovery team members in the plan test.

Goals of Testing

The immediate purpose of testing is to to provide feedback about the strategies and procedures that constitute the "fabric" of the documented recovery capability. Often this is misinterpreted to mean that tests seek to identify oversights and errors in the plan. This is only half true. Tests are carefully constructed to reinforce appropriate strategies and to demonstrate inadequacies where they exist. Testing is a vehicle for acquiring information about the plan; the results of tests are invariably used to improve the disaster recovery capability. From this perspective, one can readily support the argument that there are no test failures. Specific goals of testing include:

- To validate (and identify flaws in) plan procedures and strategies
- To obtain information about recovery strategy implementation times (to demonstrate how quickly a network recovery strategy, for example, can be implemented and networks made available for use)
- To demonstrate output performance of systems and networks operating in recovery mode or to compare performance of backup systems and networks with production systems and networks

- To demonstrate recovery plan adequacy to examiners, auditors, and management
- To adapt existing plans to encompass new requirements resulting from business, systems, networks, or personnel changes
- To familiarize recovery teams with their roles within the disaster recovery plan

To this list of immediate goals of plan testing may be added a longer range goal that is more general and fundamental. Tests provide participants with a basic set of skills for coping with a disaster, but may also be used to teach participants how to cope rationally with the irrationality of a disaster. As real disasters contain many unpredictable elements, so tests often pose unexpected challenges that participants must surmount in time-effective ways. By posing challenges that require innovative problem-solving under a deadline, testing equips participants with a skill that would be impossible to obtain in any other way—a skill that will be extremely useful in the event of an actual emergency.

Myths about Testing

Given the importance of testing, it is interesting that very little information is in print about how to test. The absence of information has, in turn, contributed to the cultivation of myths about testing that need to be dispelled before a discussion of testing techniques may be initiated.

One myth has already been alluded to: that testing is intended to prove that the disaster recovery plan works. Those who adopt this viewpoint are drawn inevitably to the conclusion that tests which demonstrate errors or oversights in plan procedures are "test failures." The position of this book—that there are no failed tests—has already been stated and justified.

A second myth about testing is described by industry experts as a "myth of realism." Specifically, the adherents of this viewpoint insist that for a test to be useful it must simulate actual disaster conditions—to the point of interrupting normal system operations! It is interesting how many otherwise sophisticated system and network professionals still shy away from plan testing because they believe that such tests require an actual shutdown of operational systems and networks in order to simulate disaster conditions. Not only is this a misperception of testing, but it also runs counter to the highest objective of disaster recovery planning: minimized downtime.

In fact, operational systems and networks are almost never shut down in a disaster recovery test. Testing is usually conducted concurrently with normal processing, whether independently or in parallel, to yield comparative data about host and backup operations. Even in so-called "mock disasters," the cessation of normal processing isn't necessary and adds nothing but heightened crisis potential to the test.

Of course, there may be another concern that leads planners to the conclusion that testing will require an interruption of normal system operations: specifically, staff shortages. Obviously, staff who are required to participate in tests cannot also be doing normal work for the company. It is, therefore, valid for some planners to ask how they will be able to both sustain normal processing and conduct a test, given the staff requirements of each activity. The answer is to conduct tests during slow periods in the production schedule, at off-peak processing times, on alternate shifts, or over weekends.

In addition to the myth of test failure and the myth of realism, yet another misconception about testing is derived from a "myth of completeness." Some companies defer testing until such time as the disaster recovery plan can be tested in its entirety. The rationale for this position is that a disaster recovery capability constitutes a complex web of interdependencies between system, network, and user recovery strategies and that testing only a subsection of the plan yields little useful information. Promulgators of the myth argue that plan strategies cannot be validated by isolated tests of specific plan procedures and that participants in such tests will not succeed in obtaining a *global view* of the recovery capability—a stated goal of testing.

In fact, as depicted elsewhere in this book, disaster recovery plans are best constructed as modular documents, with modules organized around discrete business functions and their associated automation resources. While the idea of a generalized test or mock disaster is not without merit, such a test may only be conducted profitably after procedural components of the plan have been debugged through rigorous tests of individual plan modules. In this way, the results of a generalized test will not be muddied by procedure-level errors.

A final myth about testing is the "myth of readiness"—the assumption that testing readies the company for an actual disaster. If readiness is interpreted as "armed with a baseline set of tested strategies for coping with an unplanned interruption," then the assumption is correct. However, many planners who seek a generalized test or a "realistic" test of the plan often do so because they believe that it prepares plan participants for "the real thing." This is not true.

The fact that a plan was tested last month does not mean that the company is fully prepared for a disaster this month. At best, testing familiarizes recovery personnel with their roles and provides a repertoire of recovery procedures that have been proven to work under controlled circumstances. But, as the recoveries of many companies attest, successful recovery in an actual crisis is typically the result of a combination of factors, including the existence of a tested recovery strategy, staff ingenuity and loyalty, good management, effective vendor participation, and plenty of luck. It is impossible to predict all of the consequences of a disaster. Thus, no test can fully equip a company to recover from an emergency by itself.

Test Frequency

One of the questions most often posed at meetings and conferences dealing with disaster recovery planning is, "When should we test our plan?" Given the speed with which changes occur in most company systems and networks, it is important to identify an appropriate frequency for plan testing that (1) is consistent with budget constraints and (2) provides the necessary information to keep the plan up-to-date with changes in the business it is seeking to protect.

Thus, to address the questions of "when?" and "how often?" it is first necessary to determine at what point in plan development the questions are being raised. Testing may be undertaken as part of the disaster recovery planning project—to aid in the definition and refinement of the initial disaster recovery capability; or, testing may be undertaken as part of change management—to maintain an established disaster recovery capability in a current form, accounting for changes in the business functions that are the objects of recovery strategies. Each stage has its own requirements and produces its own mindset.

Testing within the Project Plan

As this book indicates, the enterprise of disaster recovery planning is best handled as a project, controlled and managed using familiar techniques of project management. Within the disaster recovery planning project tasks are derived from project objectives, ordered according to priorities, and assigned resources, budgets, and timeframes for completion.

The completion of the overall project (and each of the task-oriented subprojects) occurs in three universal stages: analysis, implementation, and evaluation. In the analysis stage, the problem or task is defined and alternative solutions are identified. When a specific solution is selected, it is implemented. Then, to confirm that the implemented solution fulfills the objectives of the task or project, it is subjected to validation testing as part of an evaluation stage.

Within the context of the disaster recovery planning project, the evaluation stage is very important. Task outputs—embodied in procedures, strategies, or other plan deliverables—are scrutinized to ensure that tasks have been completed fully. This evaluation/validation step is required to ensure that all of the project subcomponents will ultimately work together to satisfy the main objectives of the disaster recovery planning project.

The point is that testing is a natural component of the project management approach. Testing both validates the specific outputs of project tasks and verifies that combined outputs meet general project objectives. As an integral part of the disaster recovery planning project, testing itself should be accorded *task* status and allocated budget, resources, and time. Tests may encompass any combination of plan modules and objectives and may be conducted as frequently as the planning team desires.

However, one testing strategy, which has proven its effectiveness in numerous successful recovery efforts in recent years, suggests a three year/seven-test formula that may be emulated by readers of this book. According to the formula, strategies developed by the company for recovering the mainframe operating system, critical applications, and network communications should be tested separately in three tests in the first year following plan completion. Such strategy-oriented testing provides the means for refining the strategies themselves before any generalized testing is conducted.

In year two, planners are encouraged to conduct at least two tests. One should be a retest of any year-one test (operating system, applications, or networks) in which problems were encountered. The second test may be a parallel processing test, in which a complete application processing cycle is conducted at the backup facility. This test can provide useful benchmark data that will aid planners in predicting schedule requirements and production timeframes for an actual emergency processing situation.

Project-related testing concludes in year three, with two more tests. One should consist of additional parallel testing of critical applications or parallel processing of a particular shift or schedule peak time (i.e., month-end processing) to obtain additional control data. For the final test, a mock disaster is recommended, in which operating system, applications, and networks are all brought up at the backup site.

In summary, under the project management approach, at least seven tests are conducted in the first three years after a plan is developed. The results of these tests are formative; they will accomplish the following:

- Assist the planning team in fine-tuning individual recovery strategies
- Provide baseline data on remote system performance that is required to construct a viable emergency management plan
- Ultimately, test how well the various component strategies work together to support full recovery of critical systems and networks at the mainframe backup center.

These tests should be scheduled and budgeted in the original disaster recovery planning project roadmap. They constitute the evaluation phase of the project and provide feedback that is required to complete the project mission: the development of a demonstrated disaster recovery capability.

Testing as a Change Management Function

What happens after year three? Like all projects that result in complex systems, there is the issue of maintenance and change management to consider in disaster recovery planning. Systems and networks change as the business changes over time. Change management procedures are an important part of the disaster recovery plan and work to ensure that the documented strategies, procedures, vendor relationships, and so forth are kept up-to-date.

The change management mindset is driven by events rather than project schedules. The need to retest the disaster recovery capability from the standpoint of change management may arise from several sources:

- **Major Operating System Upgrades**—Whenever there is a major upgrade in either the operating system software or utility software that affects the nucleus of the mainframe system, a test should be scheduled. For example, it is a wise idea to test following a conversion from IBM MVS/SA to MVS/XA.
- **Critical Application Software Changes**—Disaster recovery planning uses business functions to classify software applications as critical or noncritical. When new applications are added, when existing software is upgraded, or when a conversion is undertaken to move from one software vendor to another, these events should raise a flag in the change management mindset that another test is needed.
- **Major Hardware Changes**—When CPU, DASD, or other important hardware components are changed, either at the host site or at the mainframe backup facility, planners will need to retest their backup plans. Planners who use commercial mainframe backup facilities should be sure

to consult their service contracts to determine how much advance notice they will receive regarding equipment changes at the vendor site. Some vendors currently provide as much as 60 days' advance notice to ensure that their customers have time to assess the impact of the change and schedule a test.

Also, planners need to be sure to monitor changes in network hardware and configurations. Network changes are often overlooked, contributing to the observation by some industry observers that, in an actual emergency situation, network recovery is often the most time-consuming aspect of disaster recovery. When substantive changes are made to existing business networks, a new network test may be needed.

- **Personnel Changes**—The average amount of time that an individual remains in his or her job is about three years. For information systems personnel, the rate of turnover is often much higher—which is probably what prompted the observation about programmers, "Here today, gone to Maui." Following key personnel changes, job realignments, and other significant personnel-related activity, it is generally a good idea to conduct a test for training purposes.

Of course, not every change will lead automatically to a test of the disaster recovery plan. Often times, several flags accumulate before a test—which is not an inexpensive proposition—is undertaken.

In some companies, a disaster recovery planner attends all status meetings of the company data processing or information systems department. The planner reserves time on the agenda of each meeting to discuss the possible disaster recovery ramifications of new applications, equipment changes, and network reconfigurations. When the planner counts a certain number of operations-related issues with disaster recovery implications, a test is scheduled.

Types of Tests

Whether issuing from a project task or a change management *red flag*, plan testing may be conducted in a variety of ways. Types of disaster recovery tests include:

- **Modular Tests**—As the name implies, these tests are constructed to test a particular procedure or set of procedures that comprise a portion of the documented recovery plan. The purpose of such tests may be to debug the procedures, to identify oversights or errors, or to collect time statistics that may be useful in coordinating the subject procedures with other procedures.
- **Strategy Tests**—Strategy tests are used to validate entire strategies or to determine the implementation times of plan strategies. In these tests, procedures that have been subjected to modular testing previously are implemented as they might be in an actual disaster. Information may be collected about relationships between procedural tasks that was not readily apparent from modular testing or written descriptions.

- **Parallel Tests**—Parallel testing is often undertaken as part of a strategy test. In a parallel test, systems and networks are operated in parallel with normal host operations. Production loads are applied to backup systems and networks simulate the operation of the restored automation resources. The outputs of the test are (1) a validation of the recovery strategy's capability to handle the anticipated work load of restored business functions, and (2) useful data on production timeframes in backup mode that may be helpful in establishing shift schedules for a recovery operations production environment.

    Parallel tests are best structured as comparative models. In other words, tests should parallel a specific, definable object, such as a complete application production cycle, a specific production shift schedule, or a specific business function. Multiple parallel tests may be conducted over time to emulate daily processing activity, month-end processing activity, first shift production, second shift production, and so forth. By relating a parallel test to a specific object model, the most useful comparative data may be acquired.

- **Mock Disasters**—Mock disasters are tests of the emergency management plan. Generally, a mock disaster test is undertaken only after modular, strategy, and parallel tests have been conducted. In a mock disaster, a disaster scenario is articulated and recovery teams step through procedures to implement those portions of the disaster recovery plan that are required to cope with the interruption. Of course, subject systems and networks in the disaster scenario are not actually shut down. The test poses a "what if" problem and evaluates how recovery personnel react to the disaster.

All of these tests may be conducted in one of two ways: as simulations or as walk-throughs. In a simulation, testing is conducted on actual equipment—appropriate vendors are notified and a portion of the plan is implemented. In a walk-through, the entire test is conducted around a conference table, with participants indicating what they will do and when. The latter is conducted as part of preplanning for any test (except unscheduled mock disasters). The former obviously provides the greatest feedback on procedures and implementation timeframes.

HOW TO TEST

As previously stated, the first encounter most companies have with plan testing is as part of the original planning project. Testing and the development of a change management system are critical components of the project and should be treated as a project event. Figure 14.1 summarizes the major activities of plan testing and maintenance.

THE DISASTER RECOVERY PROJECT
PLAN TESTING AND MAINTENANCE

MAJOR ACTIVITIES

Develop Testing Strategy

Develop Testing Schedule

Implement Tests and Document Results

Design and Implement Change
Management System

Perform Scheduled Plan Maintenance

THE DISASTER RECOVERY
PLANNING PROJECT

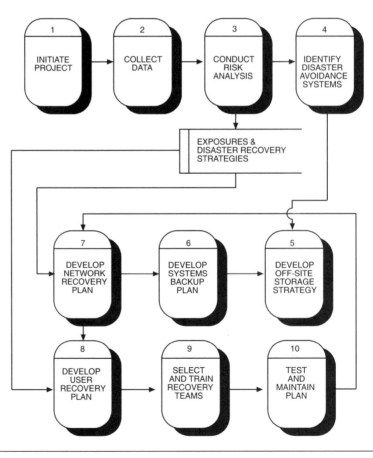

Figure 14.1 Plan testing and maintenance phase.

Develop a Testing Strategy

To conduct a test, first a strategy for the test must be established. Figure 14.2 provides a worksheet for use in articulating the general objectives of a test. This form may be used in conjunction with any of the testing options previously articulated.

In the top portion of the form, space is provided to enter up to five general test objectives. Objectives should state what will be done and what criteria will be used to measure the results. An example of this type of test objective might be: "To restore communications between the company headquarters building and the mainframe backup facility within three hours and to test communications through load simulation."

Of course, a test may also have a data collection objective. Such an objective might be structured in the following way: "To determine Accounting Department terminal response times at the user relocation facility following restoration of systems and communications links between the mainframe backup facility and the user relocation facility."

Additional test objectives may be recorded on additional strategy worksheets, with the total number of worksheets indicated at page bottom. Depending on how specifically test objectives are stated, a substantial number of objectives may be set forth for each test.

The lower portion of the worksheet is used to record the target schedule for the test, the location or locations involved in the test, and to list work unit managers and vendors who will need to be contacted to coordinate test activities. This tentative schedule will be refined as the testing strategy is developed more thoroughly.

To aid in the development of the testing strategy, a second worksheet should be used to itemize the details of each stated test objective. As shown in Figure 14.3, a test objective detail worksheet is completed for each objective articulated on the test strategy worksheet.

In the top portion of the form, the planner indicates the general test objective to which the detail worksheet pertains. The teams that will be involved in the task described by the objective are identified as well as the relevant disaster recovery plan procedures involved. Finally, the anticipated goals or results of the test are listed.

The bottom of the form is used to record the resource requirements of the test. Planners need to identify what system resources will be used, what network resources, what will be required from off-site storage, and what other forms, records or supplies will be needed.

Schedule the Test

Testing strategies are often developed as a team effort, with input from recovery team leaders, work unit managers, vendor representatives, auditors, and others. Generally, one or more members of this test-planning team will be responsible for following through with other preplanning activities, which are summarized in the checklist shown in Figure 14.4.

TESTING STRATEGY WORKSHEET

Test Objectives

INSTRUCTIONS: List the general objectives of the forthcoming test. Include criteria for evaluating test results or specific data collection goals. Attach additional pages as needed.

1 _____

_____

2 _____

_____

3 _____

_____

4 _____

_____

5 _____

_____

Test Schedule

Date and Time of Test: _____

Location(s): _____

_____

Work Units Involved: _____
(include contact data) _____

_____

_____

Vendors Involved: _____
(Include contact data) _____

_____

_____

Page 1 of _____

Figure 14.2 Testing strategy worksheet.

TEST OBJECTIVE DETAIL WORKSHEET

INSTRUCTIONS: Complete this worksheet for each general test objective indicated on the Testing Strategy Worksheet.

List the general test objective to which this worksheet pertains: _____

What teams will be involved? _____

Identify relevant disaster recovery plan procedures: _____

Describe anticipated results or goals of test: _____

Test Resource Requirements

System Resources: _____

Network Resources: _____

Off-site Stores: _____

Forms and Records: _____

Figure 14.3 Test objective detail worksheet.

TEST PRE-PLANNING CHECKLIST

NOTES

☐ Have the dates of the test been confirmed?

☐ Have test participants been notified?

☐ Have vendors been notified?

☐ Have work unit managers been notified?

☐ Has corporate management been advised of test?

☐ Have tapes and other off-site stores been checked and readied for shipment?

☐ Have travel and lodging arrangements been made for test participants?

☐ Have relevant disaster recovery procedures been copied and disseminated to test participants?

☐ Have Testing Strategy, Test Objective Detail, and Test Evaluation Worksheets been prepared?

☐ Have observers and participants been briefed about test objectives?

Prepared by: _____
Date: _____
Approved by: _____
Date: _____

Figure 14.4 Test pre-planning worksheet.

Pre-planning a test is an important component of test success. Pre-planning is used to define the logistical requirements of a test: when test time is to be scheduled at the mainframe backup facility, what personnel are to be involved in the test, what travel and lodging arrangements need to be made, and so forth.

Pre-planning also typically includes a survey of tapes and supplies stored off-site with the objective of ensuring that all backups are ready for use. While this may be viewed as "fixing" the test, it is important to verify that all required backup tapes are available and in a readable condition to ensure a smooth test. In an actual crisis, of course, pretesting of backup tapes might be impossible. Unless it is an objective of the current test to see how recovery teams cope with damaged tapes, this variable can be controlled for testing purposes. On the other hand, if inadequate or damaged backups are discovered in preplanning, this can be noted, the cause discovered, and corrective action taken to prevent faulty backups from becoming an issue in an actual emergency.

Another important aspect of preplanning is the preparatory steps that are taken before implementing the test. A planning session is an excellent forum for assembling vendor representatives and test participants for a structured walk-through of test procedures. By reviewing relevant disaster recovery procedures with everyone who will be involved, some oversights might be identified before the test begins. Backup site personnel may also be able to contribute their specific knowledge of their facility to help refine test objectives and procedures.

Planning sessions also provide a forum for brainstorming about actual disasters. Every disaster recovery procedure rests on unspoken assumptions about the conditions in which the procedure is to be initiated. Spending some time with the test group to identify and challenge some of these assumptions can help to identify other contingencies that need to be taken into account in the procedure.

At the conclusion of preplanning, the scheduled dates for the test will be confirmed; test participants, vendors, work unit managers, and senior managers will be apprised of the test; resources will be checked and verified; travel, lodging, and logistical preparations will be completed; relevant procedures will be photocopied and distributed to disaster recovery team leaders; all necessary forms will be completed; and test participants and observers will be fully briefed on what is required of them in the documentation and evaluation of test results.

Implement the Test and Document Results

At the scheduled time and date, the test is implemented. It should be noted that nearly every test includes operating system restoration, and many include the testing of selected applications or the restoration of business functions that applications support.

Tests may also focus on network restoration, increasingly of interest to disaster recovery planners in the wake of numerous communications carrier interruptions over the past five years. Network restoration poses some special problems for disaster recovery—both from the standpoint of change management and from the perspective of test requirements. Yet, network restoration is required to reconnect recovered systems to the users who make their operation purposeful.

Whatever the scope and focus of the test, an additional worksheet, shown in Figure 14.5, may be adapted to assist in test evaluation. The test coordinator should prepare the sheet by entering the general test objective at the top of the form, then summarizing test goals and criteria in the left column of the form. The form is photocopied and distributed to each individual who will be involved in the test as a participant or observer and who will be required to provide a report on results.

Following the implementation of the test, the results and observations of the evaluators will be entered in the right column of the form and at the bottom of page two. This format provides easy-to-use documentation of test results that the testing coordinator can review to identify change requirements in the plan.

In addition to worksheets, the insights of test participants and observers need to be thoroughly reviewed for what they can offer about requirements for procedural refinement or to aid in setting objectives for future tests. Every participant should be thoroughly debriefed and his or her observations included in the documentation of the test. This is often accomplished in a group setting where participants and observers are able to review test results jointly and to air their own accounts.

The results of the test are summarized in a test reporting worksheet, shown in Figure 14.6. The worksheet begins with a simple, fill-in-the-blanks description of the test, indicating when and where it was conducted and what generally was its focus. Test participants and observers are listed on the next part of the form. Each participant/observer is identified by name, organization, and role (e.g., observer, team leader, team member, vendor representative). Each participant/observer should complete evaluation forms; these should be attached to the reporting worksheet.

On the second page, test results are summarized. The first summary indicates information that was obtained from the test. Summaries may include a variety of data. For example, a modular test may have revealed that a set of procedures will not accomplish the desired objective within the required timeframe. This information must not be construed as a test failure, but as important information about the viability of a tested procedure and should be recorded on the reporting form.

On the other hand, the information gleaned from a test may be wholly objective. In a parallel processing test, for example, benchmark timings and other performance data may be taken that will provide useful input to the emergency management plan. This data, too, should be recorded.

Additional locations are provided on the form for recording specific procedural changes that have been made as a result of the test and for indicating new objectives, suggested by the current test, that will guide the development of future test strategies. As part of the test evaluation, participants noting flaws in existing test procedures are asked to attach revised versions of the procedures to their evaluations. These revisions should be reviewed by the test coordinator and, if appropriate, recorded in the disaster recovery plan.

TEST EVALUATION WORKSHEET

INSTRUCTIONS: *(To Testing Coordinator)* Copy onto this form the specific criteria and goals for one testing objective. Copy and disseminate to each test participant and observer. *(To Participant/Observer)* Beside each goal or criteria pertaining to the indicated test objective, indicate the result of the test (i.e, whether the stated goal or criteria was met in actual test performance. Summarize your observations on page two. If procedural changes are required in the disaster recovery plan, attach a marked-up copy of the existing procedure to the back of this document and submit the completed form, with all attachments, to the test coordinator.

TEST OBJECTIVE: _____
_____
_____
_____

TEST GOALS OR CRITERIA	TEST RESULT

Figure 14.5 Test evaluation worksheet.

TEST GOALS OR CRITERIA	TEST RESULT

Date of Evaluation: Evaluated by:

SUMMARY OF OBSERVATIONS (Sign and Date. Attach additional pages as needed.)

Figure 14.5 Test evaluation worksheet. (Continued)

TEST REPORTING WORKSHEET

SUMMARY OF TEST

A test of the ACME Corporation Disaster Recovery Plan was conducted from (date/time) ____/____ ____:____ to (date/time) ____/____ ____:____ at (locations) _____

_____

_____

_____

The test encompassed (check all that apply): ( ) operating systems recovery, ( ) applications recovery, ( ) network recovery, ( ) business function recovery. Specific test objectives and evaluations are attached.

TEST PARTICIPANTS The following persons participated in the test and have completed evaluations.

NAME	WORK UNIT OR COMPANY	ROLE

Figure 14.6 Test reporting worksheet.

TEST RESULTS

As a direct result of this test, the following information has been obtained:

As a result of this test, the following procedures have been revised:

Based upon the results of this test, future tests will include the following objectives:

Figure 14.6 Test reporting worksheet. (Continued)

New objectives may also spring from test evaluations. A common type of objective holds out the time required for strategy implementation as a record to be bettered in the next test. New objectives may also repeat past objectives that were not accomplished or which need to be retested because of procedural revisions.

DESIGN AND IMPLEMENT CHANGE MANAGEMENT SYSTEM

Because they provide feedback to the plan development process, tests are more than rehearsals or training exercises. They are the beginning of a process of change management that will continue after the disaster recovery planning project is complete.

It is important for a formal change management procedure to be instituted by the disaster recovery planning project prior to project termination to ensure that the capability for recovery that has been established by the project remains intact. This change management procedure may have several components, including the following.

Company Policy on Disaster Recovery Plan Maintenance

Planners should strive to safeguard the corporate investment in disaster preparedness by drafting several policy statements and urging their adoption and enforcement by senior management. Policy statements should address the need for ongoing testing of the disaster recovery plan; require work unit managers to engage in periodic reviews and updates of their own recovery plans; provide suitable funding for plan maintenance; provide suitable funding for ongoing recovery team training and company-wide awareness programs; and, mandate an annual report on all disaster preparedness activities to senior management.

Institution of a Full-time Disaster Recovery Plan Coordinator

Ideally, policy will also establish the role of disaster recovery planning coordinator as a full-time position within the company. A coordinator may work to ensure that other policy provisions are observed and can serve as an ongoing vendor interface and testing coordinator.

Formal Change Management Systems

Reporting vehicles need to be established to facilitate the communication of disaster recovery plan revisions to a coordinator, so that the company plan can be revised and maintained in an up-to-date status. Such a system is embodied in Figures 14.7 through 14.11. Following is a brief discussion of their use.

Change management depends on communications. By formalizing communications through the use of prepared forms and worksheets, those who require changes can express their requirements simply and directly. (The use of a standard change request format also facilitates the conversion of manual input into an electronic form. Thus, the forms depicted here can be used to set up a database application for change management quickly and easily.)

Figures 14.7 through 14.9 bear great similarity to one another. They are the same except for change definition selections and their titles. As indicated by the titles, 14.7 is used by work unit managers or their disaster recovery coordinators to submit change requests to the corporate disaster recovery coordinator. Figure 14.8 is used by data processing or information systems department personnel; Figure 14.9, is used by network management personnel.

In every case, the use of the form is simple. The requestor dates the request, identifies him/herself by name and work unit (or title), and provides a telephone number for follow-up.

Next, the requestor identifies the change that is prompting the request. This is done by checking the appropriate options on a list of boilerplate options. The details of the change are summarized in the space provided and the requestor is instructed to provide additional documentation that further describes the change.

In many cases, the form will be used to advise the disaster recovery coordinator of contact information changes (that is, when a contact person or alternate switches jobs, resigns, or is replaced, a new contact person is designated and telephone information is provided to the disaster recovery coordinator to maintain the notification directory). In other cases, equally simple changes will prompt the request.

In some cases, however, the changes will have far-reaching consequences for the disaster recovery plan. This is especially the case when there is a reorganization of company work units, when new systems are added, or when networks undergo substantial reengineering. These are somewhat more difficult matters for the disaster recovery coordinator to handle alone.

Thus, Figure 14.10 provides a vehicle for soliciting the assistance of technical experts within the company. This worksheet provides, again, a boilerplate vehicle for facilitating the review of change requests by knowledgeable individuals within the company (ideally, former disaster recovery planning project team members or current recovery team leaders). The top portion of the form consists of a routing slip used to direct the change request to a designated reviewer, to track the date sent, and to indicate the date that reviewer comments are requested. Also, the general subject of the change request is provided.

In the lower portion of the form, reviewer responses are captured in a checklist and summary comments format. Reviewers are asked to indicate what action, in their opinion, is required. Actions may consist of revising procedures, obtaining additional information about the change, retesting a plan module, or retesting an entire plan strategy. Obviously, more than one action category may be checked. A location is provided adjacent to the categories for the reviewer to note any special comments or considerations.

Once an action has been recommended, the reviewer (who should have access to a complete copy of the disaster recovery plan in its current version) is asked to indicate which of the listed sections of the plan he or she feels will be affected by the change. Reviewers indicate their choices by checking one or more optional categories and summarize any special considerations in the notes section.

CHANGE MANAGEMENT WORKSHEET
WORK UNIT VERSION

REQUESTOR IDENTIFICATION/PURPOSE OF REQUEST

Date of Request: _____

Requestor Name: _____

Requestor Work Unit: _____

Requestor Phone: _____

☐ Change in Work Unit Procedures
☐ Change in Personnel
☐ Change in Equipment
☐ Change in Application Software
☐ Change in Records Requirements
☐ Change in Security

☐ Change in Security Procedures
☐ Change in Notification Tree
☐ Change in Prevention Requirements
☐ Change in Communications Reqs
☐ Other Change (see below)

DESCRIPTION OF CHANGE (Attach additional pages as necessary. Indicate affected Disaster Recovery Plan procedure modules)

Please direct this Change Request to:

Jon William Toigo
Disaster Recovery Coordinator
ACME Company, Mail Stop XYZ

Figure 14.7 Change m308anagement worksheet—Work unit version.

CHANGE MANAGEMENT WORKSHEET
INFORMATION SYSTEMS VERSION

REQUESTOR IDENTIFICATION/PURPOSE OF REQUEST

Date of Request: _____

Requestor Name: _____

Requestor Title: _____

Requestor Phone: _____

☐ Change in Shift Procedures
☐ Change in Personnel
☐ Change in Equipment
☐ Change in Applicaiton Software
☐ Change in Configuration Reqs

☐ Change in Security Procedures
☐ Change in Notification Tree
☐ Change in Prevention Requirements
☐ Change in Communications Reqs
☐ Other Change (see below)

DESCRIPTION OF CHANGE (Attach additional pages as necessary. Indicate affected Disaster Recovery Plan procedure modules)

Please direct this Change Request to:

Jon William Toigo
Disaster Recovery Coordinator
ACME Company, Mail Stop XYZ

Figure 14.8 Change management worksheet—Information systems version.

CHANGE MANAGEMENT WORKSHEET
NETWORK VERSION

REQUESTOR IDENTIFICATION/PURPOSE OF REQUEST

Date of Request: _____

Requestor Name: _____

Requestor Title: _____

Requestor Phone: _____

☐ Change in Operating Procedures   ☐ Change in Security Procedures

☐ Change in Personnel   ☐ Change in Notification Tree

☐ Change in Equipment   ☐ Change in Prevention Requirements

☐ Change in Comm Facilities   ☐ Change in Configurations

☐ Change in Vendor Services   ☐ Other Change (see below)

DESCRIPTION OF CHANGE (Attach additional pages as necessary. Indicate affected Disaster Recovery Plan procedure modules)

Please direct this Change Request to:

Jon William Toigo

Disaster Recovery Coordinator

ACME Company, Mail Stop XYZ

Figure 14.9 Change management worksheet—Network version.

CHANGE MANAGEMENT RESPONSE WORKSHEET

INSTRUCTIONS: *(To Coordinator)* Attach Change Management Request Worksheet. Copy and attach relevant procedural documentation. Circulate for review and approval. *(To Reviewer)* Reveiw the attached Change Management Request. Indicate impact on plan and recommendations for revisions and testing. Return within one week to **Jon William Toigo, Disaster Recovery Plan Coordinator, ACME Company, Mail Stop XYZ.**

ROUTING SLIP

Directed to (Name of Reviewer): _____

Phone/Mail Stop: _____

Date Sent: _____

Date Return Requested: _____

Subject: _____

_____

RESULTS OF REVIEW

Actions Required: **NOTES**

- [ ] Revise Procedures
- [ ] Obtain Additional Data
- [ ] Retest Plan Module
- [ ] Retest Plan Strategy

Plan Sections Impacted: **NOTES**

- [ ] Work Unit Descriptions
- [ ] Procedure Modules
- [ ] Strategy Descriptions
- [ ] Configuration Docs
- [ ] Emergency Mgt Plan
- [ ] Vendor Directory
- [ ] Notification Directory
- [ ] Off-site Backup Plan
- [ ] Preventive Systems

Figure 14.10 Change management response worksheet.

CHANGE MANAGEMENT REQUEST LOG SHEET

REQ #	REQUESTED CHANGE	REVIEWED BY	RECOMMENDED CHANGE	CHANGE DATE	TEST?	TEST DATE

Figure 14.11 Change management request log sheet.

Tracking the submission and result of change management requests and reviews is facilitated through the use of a log sheet, such as the one provided in Figure 14.11. A database system would supplant the need for this document, since summaries can be generated as reports from such a system.

Therefore, through the use of these forms it is possible to obtain feedback directly from business work units and data processing and network management, and to evaluate the feedback to determine what needs to be done. Testing may be driven by the events described in the change requests. Test results, too, will provide useful input to maintain the plan in a current form.

Regularized Testing Schedule

Companies should test disaster recovery plans at least annually. Tests may occur more frequently if various change management considerations (previously discussed) arise.

SUMMARY

Plan testing and change management are the tools used to maintain the disaster recovery capability that has been developed by the disaster recovery planning project. Thus, the final stage of the project is also the most important. In this stage, tests are conducted to validate the baseline plan and a change management system is developed and implemented that will support the plan after the project team has disbanded.

Change management is also a candidate for automation. Planners should consider developing a change management database using any of the simple, inexpensive *flat file* packages that are currently available on the market. Be advised that more and more PC-based disaster recovery planning tools are also integrating this feature.

Checklist for Chapter

1. Identify testing alternatives and select testing strategies that will suit budget and time availability.

2. Develop test objectives.

3. Schedule testing and conduct pretesting activities.

4. Conduct the test and document results.

5. Assess test results, prepare test results, and identify opportunities for plan improvement identified in the test.

6. Develop a company policy on plan maintenance and present to management for approval and distribution.

7. Establish formal channels for information flows to the disaster recovery planning coordinator. A change management worksheet method may be useful. Respond to all requests for changes.

8. Add change management and test documentation to plan appendices for use by auditors.

Appendix A

Glossary of Terms

ACRC: The Association of Commercial Records Center.

ASCII: The American Society for Communications and Information Interchange.

ASHRAE: The American Society of Heating, Refrigeration, and Air Conditioning Engineers.

CHANNEL EXTENSION: A set of technologies designed to extend multiplexer mainframe channels to any distance, facilitating the remote operation of channel-attached devices at near-normal performance levels.

CO: A local office of a communications company.

COLD SITE: *See* shell site.

COMMUNICATIONS FACILITY: The generic term describing a combination of media and telecommunications company services used to provide a communications path between two or more locations.

DATA-FLOW DIAGRAMMING: A technique for representing the flow of data in a process from the perspective of the data itself.

DIAL BACKUP: A strategy for restoring communications when normal communications paths are interrupted.

DISASTER: An unplanned interruption of mission-critical business functions for an unacceptable period of time.

DISASTER RECOVERY PLANNING: Planning for the restoration of mission-critical business functions in the wake of a disaster. Also called Business Continuity Planning, Business Interruption Planning, Contingency Planning, etc.

EARTH STATIONS: Ground microwave receiver/transmitter facilities for use in a satellite microwave network.

EMI: Electromechanical or electromagnetic interference—an impediment to data communications.

HOT SITE: A commercial mainframe recovery facility consisting of an environmentally controlled, secure, raised floor area, and one or more mainframe computer systems that are available for company use.

MEDIA: A generic term referring to a carrier of communications signals or a means for storing information or data.

MIPS, NOT SITES: A commercial mainframe backup strategy in which customers are sold a backup capability rather than a specific hot site location. In the event of a disaster, the customer is restored at the facility of the vendor's choice.

MVS/SA, /XA, /ESA: Mainframe operating systems from International Business Machines (IBM) Corporation.

NFPA: The National Fire Protection Association.

PARTICULATE CONTAMINATION: Airborne dust and other contaminants that have a tendency to plate out over equipment surfaces and which can cause malfunctions in sensitive electronic circuits and disk-drive devices.

PC-BASED DISASTER RECOVERY PLANNING TOOLS: Microcomputer application software for use in developing and maintaining disaster recovery plans. Also called *canned plans*.

RFI: Radio Frequency Interference—an impediment to data communications.

SO: A local office of a long-distance telecommunications provider.

SHELL SITE: A mainframe recovery facility consisting of an environmentally controlled, secure, raised floor area, and necessary utility and communications connections to facilitate speedy installation of a new mainframe system.

UPS: An uninterruptible power supply. A battery backup power supply that replaces utility service in the event of an interruption. On-line models continually condition utility power, while off-line models activate within milliseconds of a detected interruption.

Appendix B

Using the Disk

INTRODUCTION

The worksheets and checklists have been prepared in the following formats:

Lotus 1-2-3 Version 2.3 or higher. These files have the extension .WK1.
Microsoft Word for Windows 2.0. These have the extension .DOC.

Please see the READ.ME write disk for description of the files.

COMPUTER REQUIREMENTS

The enclosed diskette requires an IBM PC or compatible computer with the following:

- IBM DOS or MS DOS 3.1 or later.
- A 3 1/2 inch disk drive.
- Lotus 1-2-3 Version 2.3 or higher.

Any word processor that can read Microsoft Word 2.0 for Windows.

Optional equipment includes a DOS compatible printer. If you have a different spreadsheet software package, consult your user manual for information on using Lotus files in your package. Most popular spreadsheet programs, including Microsoft Excel and Quattro, are capable of reading files formatted for Lotus. Using the index in your software manual, refer to the section in "Converting Lotus Files" or on "Loading Files from Other Programs."

INSTALLING THE DISKETTE

The enclosed diskette contains 70 individual files in a compressed format. In order to use the files, you must run the installation program for the diskette. You can install the diskette onto your computer by following these steps:

To install the files, please do the following:

1. Assuming you will be using drive A as the floppy drive for your diskette, place disk one into your floppy drive and at the A: prompt type **INSTALL**.

2. The default drive and directory settings are C:\TOIGO. If you wish to change the drive and directory names, you have the option to do so. The valuation models will be installed under the main directory.

The files are now successfully installed onto your hard drive.

USING THE FILES

Once you have installed the disk on your hard drive and made backup copies as instructed, you can begin to customize the files. The Lotus spreadsheets can be identified by the WK1 extension, and the Microsoft Word documents can be identified by the .DOC extension. To use the spreadsheets and documents, load your software programs as usual. When you are through using a file, you can save it under a new file name in order to keep the original file intact. For more information about using the spreadsheets and documents consult the appropriate software user manuals.

DISK TABLE OF CONTENTS

File Name	Title
3-3	Consultant Checklist
3-4	Deliverables Checklist
3-5	Disaster Recovery Planning Tools Evaluation Checklist
4-2 (a-g)	Contingency Planning Project Business Function Survey
4-3 (a-c)	Resource Requirements Worksheet
4-4	Terminal Usage Profile
4-5	Outage Impact Estimate Worksheet
4-7	Business Function File Cover Sheet
4-8 (a-b)	Aggregate Outage Impact Costs Worksheet
4-9 (a-c)	Preliminary Criticality Assessment Worksheet
4-10	Functional Input/Output Worksheet
4-11 (a-c)	Criticality Assessment List
5-3	Threat Ranking Worksheet
5-9	Insurance Policy Worksheet

File Name	Title
6-4 (a-c)	Disaster Prevention Inspection Worksheet
6-5	UPS Sizing Worksheet
6-8	Checklist For Reducing Environmental Contamination
6-9	Fire Prevention Checklist
6-10	Checklist for Physical Access Controls
6-11 (a-b)	Disaster Prevention
6-12 (a-c)	Disaster Prevention Bid Specification Worksheet
6-13	Preventive Systems Bid Evaluation Worksheet
6-14	Checklist For Evaluating Preventive Systems Vendor Bids
6-15	Reference Account Interview Worksheet
7-2	Records Retention Worksheet
7-3	Tape Backup Worksheet
7-4	Backup Strategy Worksheet For Small Systems
7-5	Off-site Storage Requirements Worksheets
8-1 (a-c)	Data Center Hardware Inventory
8-2 (a-c)	Data Center Software Inventory
8-4	Minimum Equipment Configuration Worksheet
10-3 (a-c)	User Recovery Center Requirements Checklist
12-3	Task Objectives/Procedures Worksheet
12-6 (a-b)	Team Definition Worksheet
12-7 (a-b)	Emergency Management Notification Directory
12-8	Departmental Notification Directory
12-9	Vendor Notification Directory
12-10	Sample Disaster Recovery Plan Table of Contents
13-2	DRS Objectives Development Worksheet
13-3	DRS Lesson Plan
13-4	DRS Training Evaluation Form
14-2	Testing Strategy Worksheet
14-3	Test Objective Detail Worksheet
14-4	Test Pre-planning Checklist
14-5 (a-b)	Test Evaluation Worksheet
14-6 (a-b)	Test Reporting Worksheet
14-7	Change Management Worksheet Work Unit Version
14-8	Change Management Worksheet Information
14-9	Change Management Worksheet Network Version
14-10	Change Management Response Worksheet
14-11	Change Management Request Log Sheet

Index